I thank Professor Saeid Jafari who read the book and advised me where I need to improve clarity so that book will penetrate the deepest core of the reader's intelligence.

Relativity Reborn

Based on Bijective Physics

$$E = mc^2 = (\rho_{E\max} - \rho_{E\min}) \cdot V$$

Amrit Srecko Sorli

Bijective Physics Institute
Slovenia

CONTENTS

1
Relativity Reborn

Relativity Theory still today is not well understood. People think it is time that is relative, but I think that it is the velocity of changes which is relative. Twin on the Moon is aging faster than his brother on the Earth because the density of the space on the Moon surface is bigger than the density of the space on the Earth surface. The density of the space is the only factor which determines the velocity of changes. Imagine you are in the spaceship which is far away from stellar objects. There you will be aging faster since the density of space there is highest. After landing on the planet with the gravity force twice bigger than planet Earth you will be aging slower than on the Earth, but you will be twice heavier. How the strong gravity is influencing your health, we still do not know, but I think for long healthy life planet Earth is the most appropriate one.

The relative velocity of your aging and in general of all changes depends only on the density of the space. In Einstein's Relativity density of space is described with the concepts of curvature, the more space is curved, less space is dense. In Relativity Reborn developed by me, time is not 4th dimension of space, but the time has only the mathematical existence. Changes are running in space only, not in time. Time is merely the duration of changes running in space. In space, time is not running on its own. Without changes, there is no time. Einstein used to say that linear time "past-present-future" is a stubborn illusion, he is right, there is no time in the universe as the fundamental physical reality in which change is running. Space is the fundamental arena of the universe and duration of changes as time is the emergent reality which in order to exist needs to be measured from the side of the observer.

This fact is intriguing and will be well explored in this book. Our measurement of the duration of an event creates duration. We are convinced in physics that we measure the duration of events which exists on its own. It is not so. We are creating a duration when we use clocks and measure it. Einstein was aware of this fact. He used to say that time is the order of events and has no independent existence apart from events. This simply means that in space is always and only NOW. The universe is in continuous change which does not run in time, it runs only in space.

With my research fellow Italian physicist Davide Fiscaletti we publish an article on this subject in Foundations of Physics back in 2015. The scientific community did not recognize the significance of this discovery. They are still seeing the universe as a system which runs in some physical time. In his book "The End of Time" British physicist Julian Barbour is also denying the existence of time as some physical reality. The book with the same title O published back in 1990 in Slovene language. In his recent book "Time Reborn" American physicist Lee Smolin tries to convince us that time has some physical existence, but he does not mention a single data in his book which would confirm that time has a physical existence. Full acceptance of the fact that time has only the mathematical existence is the needed step for physics and cosmology progress.

In our daily life, we experience the flow of changes in the frame of linear psychological time which is only on the human mind. There is no such linear time in the physical universe. This insight was the first step in my development of Relativity where universal space has the origin in the multidimensional vacuum. In this book, I will use the term "vacuum" instead of the term "space", because in physics today we know, that universal space has the physical origin in the vacuum which is primordial non-created energy of the universe.

The vacuum which represents 95% of the energy in the universe has no entropy. Only 5% of the universal energy in the form of matter has entropy. We have been in the physics of the 20th century giving too much importance to the second law of thermodynamics and we have been giving too much importance also to the time. The universe is governed by the first law of thermodynamics which says, energy cannot be created or destroyed, it can only change its form.

The vacuum is dynamic in the sense that interacts with the physical objects. It diminishes the density on the surface of a given physical object exactly for the amount of the energy of the object. This can be expressed in the formula below which somehow I develop after 5 years of continuous research:

$$\frac{E}{c^2} = m = \left(\rho_{\max} - \rho_{\min} \right) \cdot V$$

This formula is the basis for the development of the Relativity where mass-energy equivalence principle is extended on the vacuum where vacuum is a kind of superfluid which has a given density. Model of the vacuum as a superfluid is developed by Russian physicist Valeriy I. Sbitnev. His excellent articles on the vacuum you can find on arXiv. The article where I extend mass-energy equivalence on the vacuum was published in August 2019 in the journal Scientific Reports (Nature&Springer). Article is entitled: "Mass–Energy Equivalence Extension onto a Superfluid Quantum Vacuum". The formula above can also be written in the following form:

$$E = mc^2 = \left(\rho_{E\max} - \rho_{E\min} \right) \cdot V$$

where $\rho_{E\max}$ is energy density of the vacuum in interstellar space and $\rho_{E\min}$ is the energy density of the vacuum on the surface of a given stellar object or massive object. This formula suggests that vacuum is primordial non-created energy of the universe which has variable energy density. Vacuum is not a kind of "superfluid" which has variable density; it is primordial energy of the universe which has variable energy density. I prefer this form of formula.

Dear reader, this brief presentation is your companion throughout the entire book. Keep in your mind that in the universe is always NOW and that space is the primordial non-created energy of the universe. Reading this book with this mindset will be of immense help into your understanding of Relativity and also in comprehending of Einstein's "Unified Field Theory".

Do not read this book with your belief that space-time is a fundamental arena of the universe. You will end with the headache. In order to progress physics, we have to step out of our psychological time and enter the vastness of NOW which is the source of true intelligence.

2
Improvement of Einstein's Mathematical Tricks

By the end of 19th-century ether was thrown out of physics. Michelson-Morley experiment has given the null result and the model of ether which has been serving physics for two centuries was demolished. After ether was abandoned the question has arisen, how the photon can move in the universal space without a medium. Einstein has solved the question by a very superficial explanation, namely, that photon can move in an "empty space". I will show in this book that the idea of "empty space" is Einstein's big theoretical mistake.

In order to progress physics, at the beginning of the 20th-century, there was an unsolved puzzle, namely, physicists discover that light has the same speed in all inertial systems for all observers. In 1905 Einstein has published his Special Theory of Relativity where the problem was mathematically solved with the introduction of "length contraction" and "time dilatation. Time has become the 4th dimension of space without having single experimental evidence for such interpretation of time.

With Italian physicist Davide Fiscaletti, we have been developing SR in a 3D Euclidean space, with Galilean transformation and Selleri formula for time, without length contraction and time dilation. Several articles were published in Physics Essays and some other journals. The fact is that in the universe time is not relative; relative is the velocity of the material changes, rate of clocks included. In the universe, there is no "internal observer", no "external observer", no "proper time", no "coordinate time", which all are Einstein's mathematical tricks in order to describe a constancy of light speed for all observers.

Keeping ether as the physical origin of the universal space the constancy of light is elegantly preserved. The source of light is in the ether (which today we call "super-fluid quantum vacuum") all observers move in the same ether. Photon is the vibration of the ether and has the same speed for all observers. The speed of light is invariant for all observers and obeys Doppler effect. This description is in accord with all the experimental data we have in physics. It is time we abandon Einstein's mathematical tricks and build physics which is based on the observation and experiment.

Einstein has developed the mathematical trick of length contraction in General Theory of Relativity, where hypothetical gravitational wave should shrink or dilate space-time. Time has only the mathematical existence, so it cannot shrink or dilate. I published together with Italian physicist Davide Fiscaletti an article on the time as the mathematical parameter of material change, i.e. motion in Foundations of Physics back in 2015. The article is titled: "Perspectives of the Numerical Order of Material Changes in Timeless Approaches in Physics«.

What happened in LIGO is that gravitational waves are changing the variable density of the vacuum. Change of variable density of vacuum is changing its permeability and permittivity which is minimally changing the light speed. In LIGO was measured that light beam when passing the interferometers is changing minimally the speed. The interpretation of these results is wrong. A gravitational wave is not shrinking and dilating space-time since space-time does not exist. The vacuum is the origin of universal space and in the vacuum is always NOW.

For Einstein, the NOW remained the puzzle for his entire life. He did not manage to integrate NOW into physics. I happily "divorced" space and time and this separation allows the development of Dynamic Vacuum Relativity (DVR) which integrates Relativity, Quantum electrodynamics, and Higgs mechanism.

3
Mathematics should be the Servant of Physics

Today, physics has several prominent models: quantum electrodynamics (QED), the General Theory of Relativity (GR), Quantum Physics, and the Higgs mechanism. The crisis of today's physics is that these models are not connected in a satisfying way. Each of these models has a well-developed mathematical description and works well in its own domain. Because the reality is one, we are looking forward to connecting these partial pictures in one general picture that will embrace the entire physical existence. Such a picture would mean that we are surpassing the crisis of today's physics.

For several years, I have been trying to figure out how to unify these main models. It is not an easy job and many others are working on it. Finally, I decided to help myself with set theory. We can assume that every element in the physical reality is an element of the set X, which is the universe. To build a unifying picture of the universe we have to ensure that every element in the universe will have exactly one corresponding element in the model of the universe, which is set Y.

Adopting this methodology, I started building the model of the universe on my own without much consideration for the established knowledge of physics; my proposition was that in physics there are some cardinal mistakes which are preventing the unification of the leading models of today's physics.

My mind was never in tune with the idea of space-time being the fundamental arena of the universe. Applying a bijective methodology, it was proved that the space-time model has no physical existence. Time is only the mathematical parameter of material change, i.e. motion in space. Space is not 'empty', space is a kind of energy which we are not familiar with yet. In the QED model, we are using the electromagnetic vacuum as the physical ground for electromagnetism. I assume that it is right to replace space-time as the fundamental arena with the electromagnetic vacuum in which time is only a mathematical parameter.

The next step was integrating GR and QED. My idea was that the curvature of space in GR is related to the density of the electromagnetic vacuum. The more space is curved, the less dense is the vacuum. This model is a powerful tool in the description of the gravitational force, which is not acting directly between two physical objects; it is acting indirectly via the variable density of the electromagnetic vacuum. A given physical object with mass m is diminishing the density of the vacuum exactly for the amount of its energy, E. This diminished density is generating the inertial mass and gravitational mass of the given physical object.

Quite amazing: with the introduction of the variable density of the electromagnetic vacuum, we can unify QED, GR, and the Higgs mechanism. Even more, we can solve the 'enigma' of dark energy, which is the energy of the electromagnetic vacuum itself. In today's physics, the Standard Model is the main player, which also gets most of the financial support money for research. The standard model establishment will fight my idea with all possible force. Bijective physics implies the end of the Standard Model supremacy.

I am highly suspicious of the CERN research methodology which has discovered the 'God particle'. In my opinion, the Higgs boson is nothing more than the momentary flux of energy which is released here and there in proton collisions; nothing spectacular and with no possibility for any technological application. By investing 20% of the money that CERN has used over the last 10 years, we would already have discovered antigravity and the extraction of energy directly from the electromagnetic vacuum.

Einstein's idea of the four-dimensionality of space is intriguing for me. This would mean that the electromagnetic vacuum is the 4-dimensional (4-D) reality. The photon is the wave of 4-D vacuum and because of this; we cannot fully grasp it with 3-D apparatus. I am in favor of the idea that the proton is a 4-D vacuum vortex and has no constitutive parts. With acceleration, protons gain their relativistic energy, which is the energy of the vacuum. By smashing protons, we are getting various sparkles; one of them is the 'Higgs boson'. Quantum physics should treat all elementary particles as 4-D structures of the electromagnetic vacuum.

Bijective physics is strictly against the novelty of today physics, namely, 'playing with mathematics' produces new models. For example, in theoretical physics, negative mass is a matter whose mass is of opposite sign to the mass of normal matter, e.g. −1 kg.

The idea of negative mass is introduced by British physicist Jamie Farnes. Back in December 2018, he published an article titled »A unifying theory of dark energy and dark matter: Negative masses and matter creation within a modified ΛCDM framework« in Astronomy&Astrophysics journal. His article is mathematically correct, but it does not reflect physical reality because »negative mass« does not exist in the physical universe. Negative mass model is a school example of how theoretical physics went on the wrong path. Today you will publish an article in the best journal of physics just by developing the model which is mathematically well described. How much your model corresponds to the physical reality nobody is bothered. This is deepening the crisis of today physics. The way out is Bijective physics.

The idea of negative mass is a clear sign that physics is in danger of becoming a philosophy. And philosophy is not far from psychotherapy. We have many schools of psychotherapy and we do not know exactly which one really works. This should not be the case with physics. The bijective research methodology is the guarantee that physics will develop in the right direction.

4

Introduction to Bijective Physics

For more than 100 years physics is building the models in which reality is squeezed in different unnatural forms. The result of this "forced" methodology is that we are now discovering things which do not exist. It is time to rebuild physics accordingly to the shapes of reality.

Bijective physics is using bijective research methodology in order to build a "two-way" bridge between models of reality and reality itself. In Bijective physics, every single element in a given model corresponds to exactly one element in physical reality. A given element of the reality   and a given correspondent element of the model are related by bijective function:

$$f : X_R \to Y_M$$

In today physics often happens that a given model is build on purely theoretical predictions without being observed directly by senses or indirectly by instruments. Such an example is the idea that all elementary particles are mass-less and that some field exists which gives mass to the particles. The idea was born as the result of Super-symmetry model SUSY which predicts that every elementary particle should have its "super-partner" particle. Higgs mechanism is mathematically correct, but it has no basis in human perception (direct perception or indirect perception by instruments). It is not in accord with the "mass-energy equivalence principle" and formula $E=mc^2$ which clearly confirms that mass of the given elementary particle is equal the content of its energy, so mass has the origin in the energy of a given particle.

The idea that all particles are "mass-less" is false and has lead to the extremely complicated model of the Higgs mechanism, which cannot answer the simple question, as for example: About 99% of proton mass consist of gluons. Higgs field does not interact with the gluons. How Higgs field can give mass to the proton?

Bijective Physics offers a model where Higgs potential is elegantly described without the variable density of the vacuum. The model is presented in the next chapter.

Next example of the model which has no bijective correspondence with reality is the idea of "negative mass" in order to solve the enigma of dark energy and dark matter. Such mathematical proposals are highly complicated physics and do not bring any constructive solutions. The term "negative mass" is mathematically correct, but it has no meaning and should not be introduced in physics. It does not pass bijective analysis. The same is valid for the model of "negative energy". The idea was introduced by Hawking in his book The Brief History of Time, where he has presented energy of matter as "positive energy" and energy of gravity which is energy of space as "negative energy"; their sum in the universe is always zero:

$$E_m + (-E_g) = 0$$

Hawking suggests that in the "inflation period these two energies are getting multiplied. He explained this with the fact that the sum of positive natural number and the came negative natural number is always zero. It is right that (1) + (-1) = 0, 2 + (-2) = 0, and so on. But this does not mean that energy can be produced in the universe on the basis of this equation.

A 100 years old model of cosmology and physics, namely, that space is "empty" is a false idea. We cannot bring the model of Euclidean mathematical space in physics and believe universal space has only "geometrical properties". Such a model has created the unsolvable problems in physics, which bijective physics is solving elegantly.

5
Dynamic Vacuum Relativity (DVR)

The superfluid quantum vacuum model is replacing space–time as the fundamental arena of the universe. In the superfluid vacuum (from now on 'vacuum') time is the numerical sequential order of material changes, i.e. motion running in a vacuum. The vacuum is timeless in the sense that time is not its fourth dimension. The vacuum is the direct information medium of entanglement regarding Einstein-Podolski-Rosen type (EPR-type) experiments. Back in 2017, we published with my Italian research fellow Davide Fiscaletti an article on this subject in Quantum Studies: Mathematics and Foundations. Article is entitled: "Searching for an adequate relation between time and entanglement". The message of this article is the following: 'Today, mainstream science considers that the observer and all observed physical phenomena exist in time and space. Nonetheless, recent research shows that the time measured with clocks is merely a mathematical parameter of a material change, i.e. motion which runs in space. In this picture, the existence of past, present, and future is merely a mathematical one. As regards EPR-type experiments, observer and observed phenomena exist only in space which originates from a fundamental quantum vacuum which is an immediate medium of quantum entanglement'.

The article is not acknowledged yet by the scientific community. In recent scientific literature, we have a lot of articles on the subject of entanglement with no really convincing results because all the authors see entanglement in the model of space-time which does not exist. A classical example is the article of Dr. Ekaterina Moreva and collaborators titled: "Time from quantum entanglement: An experimental illustration". The article was published in prestigious journal "Physical Review A". Authors claim that an "internal" observer that becomes correlated with the clock photon sees the other system evolve, while an "external" observer that only observes global properties of the two photons can prove it is static. The whole idea is false, because all physical events, the rate of clocks included, are valid for all observers. GPS system proves this. Einstein's "external observer", "internal observer", "proper time", coordinate time" is all nonexistent in the physical universe and we cannot build physics on them.

The formula for the variable density of the vacuum is defined by the mass and volume of a given stellar object. Let us imagine an ideal stellar object with mass that is 93 billion light-years distant from other stellar objects, which is the diameter of today's observable universe. At the distance of 93 billion light-years from this ideal stellar object, we can assume that the density of the vacuum has a maximum value ρ_{max} . On a stellar object's surface, the density of the vacuum is at the minimum (ρ_{min}). The difference between these two densities is $\Delta\rho$. A given ideal stellar object has diminishing density of vacuum on its surface exactly for the amount of its mass m . Considering that inertial mass m_i and gravitational mass m_g are proportional to the mass as the amount of energy which is incorporated in a given stellar object, we can write the following equation:

$$m_i = m_g = m = (\rho_{max} - \rho_{min}) \cdot V \ , (1)$$

where V is the volume of the physical object. The vacuum density difference $\Delta\rho$ is the source of permanent vacuum fluctuations in the direction from ρ_{max} towards ρ_{min}. Inertial mass m_i and gravitational mass m_g of a given ideal stellar object both have their origin in these vacuum fluctuations (from now on **VF**).

Density of the vacuum on the surface and of the stellar object we can calculate with the rearranging the equation (1) as follows:

$$\rho_{min} = \rho_{max} - \frac{m}{V}$$

where ρ_{min} is the density of the vacuum on the surface of the stellar object.

Density of the vacuum at the distance d from the stellar object surface is following:

$$\rho_{min} = \rho_{max} - \frac{3m}{4\pi \cdot (r+d)^3}$$

where r is radius of the stellar object. When d is going towards the infinity, ρ_{min} goes towards ρ_{max}.

Inside the stellar object, the density of the vacuum ρ is increasing by the Newton shell theorem. Newton shell theorem says that gravity force on a physical object at the given point below the stellar object surface is equal to the force which acts on the physical object as the shell above the physical object would not exist.

At the distance r from the centre, the density of the vacuum ρ, is the following:

$$\rho = \rho_{min} + \frac{3m_1}{4\pi \cdot r^3},(4)$$

where m_1 is the mass of the stellar object inside the shell and r is the radius of the shell. By increasing the vacuum density towards the centre of the stellar object, vacuum fluctuations are moving from the centre to the surface of the stellar object. Inside physical objects, we have two movements of vacuum fluctuations. One is from above towards the centre: $VF_{\leftarrow}$. The other is from the centre to the surface: $VF_{\rightarrow}$. These vacuum fluctuations are characteristic from the macro scale of the stellar objects to the micro scale of the proton.

Vacuum fluctuations, binding and repulsive pressure of the proton, Casimir forces, and van der Waal Forces

Recent research bone by Burked and others, published in Nature back in 2018 (*Nature* **557**, 396–399), confirms a strong repulsive pressure near the centre of the proton (up to 0.6 femtometres) and a binding pressure at greater distances. In the model presented here vacuum fluctuations $VF_\rightarrow$ create binding pressure of the proton. Vacuum fluctuations $VF_\leftarrow$ create repulsive pressure of the proton.

Different authors are differently describing the Casimir effect. Nikolic says following in his article published in Physics Letters (*Physics Letters B* **761**, 197–202, 2016): "The Casimir force is widely viewed as a force that originates from the vacuum energy, which is a view especially popular in the community of high-energy physicists. Another view, more popular in the condensed-matter community, is that Casimir force has the same physical origin as van der Waals force, which does not depend on energy of the vacuum. From a practical perspective, the two points of view appear as two complementary approaches, each with its advantages and disadvantages".

In the model presented here, vacuum fluctuations $VF_\rightarrow$ are the origin of the Casimir effect when we have attraction force between plates. Repulsive forces between the plates are originated by vacuum fluctuations $VF_\leftarrow$. Also, van der Waals force can be described by vacuum fluctuations.

Recent research fone by Klimchitskaya, G. L. and others titled »Casimir and van der Waals forces: Advances and problems« (Proceedings of Peter the Great St. Petersburg Polytechnic

Univercity, N1 **517**, 41–65, 2015), suggests there is no difference between Kasimir and van der Waal forces: "In fact, there are no two different forces, van der Waals and Casimir. The van der Waals force is a subdivision of dispersion forces acting at very short separations up to a few nanometers, where the effect of relativistic retardation is very small and can be neglected. As to the Casimir force, it is a subdivision of dispersion forces which acts at larger separation distances, where the effect of relativistic retardation should be taken into account. It is evident that there is some transition region between the two kinds of dispersion forces".

Vacuum fluctuations are the origin of gravity

Gravity force from the macro- to the microscale (proton) is the result of vacuum fluctuations **VF**. The gravity force between physical objects is immediate. It does not require time and motion as is the case with photon propagation in space. The gravity force F_g between an object with mass m_1 and an object with mass m_2 is expressed by the following equation:

$$F_g = \frac{m_{g1} \cdot m_{g2} \cdot G}{r^2},$$

In General Theory of Relativity gravity is carried by the curvature of space. A given physical object is curving the space and curvature of space is generating gravity. In the model presented here vacuum is the physical origin of space. The variable density of space is generating vacuum fluctuations **VF** which are generating gravity. In both models, gravity is the result of properties of space (geometrical and physical properties) and is not acting directly between two physical objects.

NASA research confirms universal space is 'flat', it has a Euclidean shape, with only a 0.4% margin of error [9]. NASA results are suggesting that curvature of space in General Relativity is only the mathematical description of its actual density, which means the density of vacuum which is the physical origin of space. The curvature of space has only a mathematical existence and cannot carry gravity. The physical origins of gravity are vacuum fluctuations.

The idea of quantum gravity theory is that gravity is carried by some quanta called "graviton" for which we do not have a single observation. Yuan K. H. says this clearly in his article "An Underlying Theory for Gravity" published on arXiv, back in 2012: "Quantum gravity has been conjectured for almost 80 years since the introduction of the graviton. It is commonly believed that gravity is a fundamental interaction and as such, it would obey quantization similar to electrodynamics. However, it is significant to point out that there is not a single observational evidence so far showing the need of a quantum theory of gravity". In the model presented in this article gravity is not quantum phenomena. Gravity is the result of vacuum fluctuations $VF_{\rightarrow}$ which are generated by the presence of a given physical object.

Vacuum fluctuations and gravitational potential

The strength of vacuum fluctuations **VF** which generate inertia and gravity we express by gravitational potential. Gravitational potential V depends on the difference between the density of the vacuum in interstellar space and the density of the vacuum at the given point T.

At an infinite distance from a given stellar object, the gravitational potential V is zero, the density of the vacuum has its maximum value. At point T above the stellar object, the gravitational potential V value is calculated by formula below:

$$V = -\frac{GM}{r}$$

where r is the distance from the centre of the stellar object, G is the gravitational constant and M is the mass of the stellar object. On the stellar object surface at point T, gravitational potential is calculated by formula above. Inside the stellar object at point T, we calculate the gravitational potential with the formula below:

$$V = -GM\left(\frac{3R^2 - r^2}{2R^3}\right)$$

where R is the distance from the centre to the point T. In the centre of the stellar object at the point T, r is zero, R is zero and the gravitational potential V is zero too.

Mass–energy equivalence extension onto the vacuum

The gravity force in space is always there in the form of vacuum fluctuations. If there is no physical object at the points at the given point in space, gravity forces have no physical object to act upon, but vacuum fluctuations are there. Both inertia and gravity are the result of vacuum fluctuations, which have their origin in the variable density of the vacuum.

The curvature of space in General Relativity is a mathematical description of the variable density of the vacuum. The more space is curved, the less dense is the vacuum. Most of the universal space has a maximum value of vacuum density, ρ_{max}. The vacuum density is decreasing in the areas with galaxies where universal space is flat too. The vacuum is the physical origin of the universal space, which means we can see a variable density of vacuum as an actual variable density of space. There is a fundamental dynamics between a given physical object with mass m and variable energy of space which we can describe with the following equation:

$$\frac{E}{c^2} = m = (\rho_{max} - \rho_{min}) \cdot V \, ,$$

where E is the energy of the vacuum that is incorporated in a given physical object, m is the mass of the object, ρ_{max}

is the density of space in the intergalactic area, ρ_{min} is the density of space on the surface of the physical object and V is the volume of a given physical object. This fundamental dynamics is the origin of mass–energy equivalence, inertia and gravity.

For relativistic particles, as for example a relativistic proton, the relativistic energy is the following:

$$E = \gamma \cdot m_0 c^2 = (\rho_{max} - \rho_{min\,R}) \cdot V \cdot c^2,$$

where E is the proton relativistic energy, γ is the Lorentz factor, m_0 is the proton rest mass and $\rho_{min\,R}$ is the density of the vacuum at the relativistic proton surface. The proton, when accelerated, is interacting with the vacuum and additionally incorporating some of its energy.

Italian physicist Marco Fedi has developed a model of the vacuum as a shear-thickening (dilatant) fluid (the Newtonian fluid). His model is presented in the article titled "Relativistic mass due to a dilatant vacuum leads to a quantum reformulation of the relativistic kinetic energy« published in Canadian Journal of Physics in May 2019. In his model relativistic energy of the proton can be seen as accelerated proton thickens the vacuum ahead of it.

If the accelerated proton is absorbing the vacuum energy

or is thickening the vacuum energy ahead of it remains an open question for now. Important is that both models see the relativistic energy of the proton as the energy of the vacuum which is absorbed or is thickening ahead of the proton. Proton does not gain its relativistic energy because of the motion in an empty space. Proton relativistic energy is vacuum energy which is interacting with the proton due to its motion in a vacuum.

The density of the vacuum on the black hole surface, neutron star surface and proton surface

The density of the vacuum ρ_{min} on the surface of a black hole with the mass of the Sun and radius of 3000 metres is the following:

$$\rho_{min} = \rho_{max} - \frac{1.989 \cdot 10^{30} kg}{1{,}131 \cdot 10^{11} m^3}$$

$$\rho_{min} = \rho_{max} - 1.759 \cdot 10^{19} \, kg/m3$$

The density of the vacuum ρ_{min} on the surface of planet Earth is given by the following:

$$\rho_{min} = \rho_{max} - \frac{5.972 \cdot 10^{24} kg}{1{,}083 \cdot 10^{21} m^3}$$

$$\rho_{min} = \rho_{max} - 5.514 \cdot 10^{3} \, kg/m^3$$

The density of the vacuum ρ_{min} on the surface of the proton is given by the following:

$$\rho_{min} = \rho_{max} - \frac{1.672 \cdot 10^{-27} \, kg}{2.5 \cdot 10^{-45} \, m^3}$$

$$\rho_{min} = \rho_{max} - 6.688 \cdot 10^{17} \, kg/m3$$

The density of the vacuum on the surface of a neutron star is $\rho_{min} = \rho_{max} - 2.0 \cdot 10^{26} \, kg/km^3$, which is $\rho_{min} = \rho_{max} - 2.0 \cdot 10^{17} \, kg/m^3$.

Regarding the maximum density ρ_{max} which is constant, the density of the vacuum ρ_{min} on the surface of the black hole is of the order $- 10^{19}$. Regarding the maximum density ρ_{max} , the density of the vacuum ρ_{min} on the surface of the proton is of the order $- 10^{17}$. Regarding the maximum density ρ_{max} , the density of the vacuum ρ_{min} on the surface of the neutron star is of the order $- 10^{17}$. Regarding the maximum density ρ_{max} , the density of the vacuum ρ_{min} on the surface of the planet Earth is of the order $- 10^3$.

Recent research results are that the average peak pressure near the centre of the proton is about 10^{35} pascals, which exceeds the pressure estimated for the most densely packed known objects in the universe, neutron stars. The calculations above confirm minimal density of the vacuum on the proton surface is $\rho_{min} = \rho_{max} - 6.688 \cdot 10^{17} \, kg/m^3$. Minimal density of the vacuum on the neutron star surface is $\rho_{min} = \rho_{max} - 2.0 \cdot 10^{17} \, kg/m^3$. Density of the vacuum on the proton surface is smaller from the density of the vacuum on a neutron star surface. That is why the peak pressure near the centre of the proton exceeds the peak pressure in neutron stars.

The density of the vacuum on the surface of a proton is $\rho_{min} = \rho_{max} - 6.688 \cdot 10^{17} \, kg / m3$. The density of the vacuum on the surface of a black hole is $\rho_{min} = \rho_{max} - 1.759 \cdot 10^{19} \, kg / m3$. On the surface of a black hole, the density of the vacuum is too low to keep a proton stable. Protons are falling apart and disintegrating back into the energy of the vacuum. This reduces the mass and the energy of the black holes.

Steven Hawking predicted that the mass and energy of a black hole are diminishing because of thermal radiation, also known as black hole evaporation. A recent article has reported the observation of quantum Hawking radiation in an analogue black hole. Another recent article raises severe doubts about the observation of Hawking radiation.

The proton rest mass is $m_0 = 1.672 \cdot 10^{-27} \, kg$. In an accelerator, the proton relativistic energy reaches in terms of rest mass m_0 a value of $E = m_0 \cdot c^2 \cdot 7460$. When this relativistic energy would be considered as mass, the relativistic proton would become a mini black hole. The relativistic energy of the accelerated proton is the energy of the vacuum, which is additionally integrated into the proton. We cannot consider this energy as a mass which would diminish the density of the vacuum because it is on the black hole surface. This means that the existence of mini black holes predicted by Stephen Hawking is questionable. Voyager data excludes the existence of mini black holes.

Einstein tensor and Variable Density of Vacuum

In General Relativity curvature of space is expressed with Einstein tensor:

$$G_{\mu\nu} = \kappa \cdot T_{\mu\nu} \qquad (1)$$

$$\text{in units: } \frac{1}{n^2} = \frac{m}{kg} \cdot \frac{kg}{m^3}$$

where curvature tensor $G_{\mu\nu}$ measures curvature of space, $T_{\mu\nu}$ is stress-energy tensor, κ is Einstein constant: $\kappa = 1,866 \cdot 10^{-26} \, mkg^{-1}$. We develop heuristic formula:

$$\vec{r}_{\mu\nu} = \kappa \cdot (\rho_{max} - \rho_{min}) \qquad (2) ,$$

where ρ_{min} is density of the vacuum on the surface of a given stellar object, ρ_{max} is density of the vacuum in interstellar space. When ρ_{min} is close to the ρ_{max} , curvature tensor $G_{\mu\nu}$ is close to zero.

We will now calculate curvature tensor $G_{\mu\nu}$ on Black hole surface, on the Earth surface, neutron star surface and proton surface. As ρ_{max} is the same in all calculation, wewill leave it out. The density of the vacuum ρ_{min} on the surface of a black hole with the mass of the Sun and radius of 3000 metres is the following:

$$\rho_{min} = \rho_{max} -1{,}759 \cdot 10^{19} \, kg/m3 \qquad \textbf{black hole}$$

$$G_{\mu\nu} = 1{,}866 \cdot 10^{-26} \, mkg^{-1} \cdot -1{,}759 \cdot 10^{19} \, kgm^{-3}$$

$$G_{\mu\nu} = -3{,}28 \cdot 10^{-7} \, m^{-2}$$

The density of the vacuum on the surface of a neutron star is the following:

$$\rho_{min} = \rho_{max} -2{,}0 \cdot 10^{17} \, kg/m^3 \qquad \textbf{neutron star}$$

$$G_{\mu\nu} = 1{,}866 \cdot 10^{-26} \, mkg^{-1} \cdot -2{,}0 \cdot 10^{17} \, kgm^{-3}$$

$$G_{\mu\nu} = -3{,}73 \cdot 10^{-9} \, m^{-2}$$

The density of the vacuum ρ_{min} on the surface of the proton is the following:

$$\rho_{min} = \rho_{max} -6{,}688 \cdot 10^{17} \, kg/m3 \qquad \textbf{proton}$$

$$G_{\mu\nu} = 1{,}866 \cdot 10^{-26} \, mkg^{-1} \cdot -6{,}688 \cdot 10^{17} \, kgm^{-3}$$

$$G_{\mu\nu} = -12{,}48 \cdot 10^{-9} \, m^{-2}$$

The density of the vacuum ρ_{min} on the surface of planet Earth is the following:

$$\rho_{min} = \rho_{max} - 5.514 \cdot 10^3 \, kg/m^3 \qquad \textbf{Earth}$$

$$G_{\mu\nu} = 1{,}866 \cdot 10^{-26} \, mkg^{-1} \cdot -5{,}514 \cdot 10^3 \, kgm^{-3}$$

$$G_{\mu\nu} = -10{,}29 \cdot 10^{-23} \, m^{-2}$$

Black hole:

$$\rho_{min} = \rho_{max} - 1{,}759 \cdot 10^{19} \, kgm^{-3} \qquad\qquad G_{\mu\nu} = -3{,}28 \cdot 10^{-7} \, m^{-2}$$

Neutron star:

$$\rho_{min} = \rho_{max} - 2{,}0 \cdot 10^{17} \, kgm^{-3} \qquad\qquad G_{\mu\nu} = -3{,}73 \cdot 10^{-9} \, m^{-2}$$

Proton:

$$\rho_{min} = \rho_{max} - 2{,}0 \cdot 10^{17} \, kgm^{-3} \qquad\qquad G_{\mu\nu} = -3{,}73 \cdot 10^{-9} \, m^{-2}$$

Earth:

$$\rho_{min} = \rho_{max} - 2{,}0 \cdot 10^{17} \, kgm^{-3} \qquad\qquad G_{\mu\nu} = -3{,}73 \cdot 10^{-9} \, m^{-2}$$

Proton has on its surface smaller density of the vacuum and bigger value of the curvature as neutron star and Earth. The smaller density of the vacuum of the proton regarding the neutron star corresponds with the recent research which has confirmed that the peak pressure near the centre of the proton exceeds the peak pressure in neutron stars. The article about this important discovery was published by Burkert and Elouadrhiri in Nature back in 2018. The peak pressure in the proton and in the neutron star has origin in the variable density of the vacuum which is smaller in the proton than in the neutron star.

Einstein cosmological constant and Variable Density of the Vacuum

In order to fit into the stationary model universe, Einstein has added into his tensor cosmological constant $\Lambda g_{\mu\nu}$. In Dynamic Vacuum Relativity (**DVR**) cosmological constant $\Lambda g_{\mu\nu}$ represents the fundamental content of energy in the given volume of the universal space. In **DVR** universal space is not empty and deprived of the physical properties. Energy of the vacuum is the physical origin of space. Cosmological constant $\Lambda g_{\mu\nu}$ represents this ground energy of the universe. In **DVR** cosmological constant $\Lambda g_{\mu\nu}$ represents the dark energy and the dark matter:

$$\Lambda g_{\mu\nu} \approx dark.energy + dark.matter$$

As we can see on the figure below, the cosmological constant $\Lambda g_{\mu\nu}$ is equal in the entire universe and does not depend on the variable density of the vacuum.

In interstellar space $\rho_{max} = \rho_{min}$, which means the curvature of space is zero:

$$G_{\mu\nu} = \kappa \cdot (\rho_{max} - \rho_{max}) \, ,$$
$$G_{\mu\nu} = 0 \, . \, .$$

In the proximity of the stellar object, a cosmological constant remains unchanged despite the density of the vacuum has diminished. This is because the diminished density of the vacuum is corresponding exactly to the mass of the given stellar object with the mass m and volume V:

$$\rho_{max} = \rho_{min} + \frac{m}{V} \, ,$$

where ρ_{min} is the density of the vacuum on the surface of the stellar object and V is the volume of the stellar object. Formula above confirms that the average density of space which represents Enstein cosmological constant remains unchanged because also the mass of the given stellar object is adding to the averge density of the space.

Variable vacuum density and variable rate of clocks

What is the value of vacuum density ρ_{max} (which when multiplied by c^2 becomes vacuum energy density) is a big dispute in today's physics: 'The theoretical vacuum energy density estimated on the basis of the Standard Model of particle physics and very general quantum assumptions is 59 to 123 orders of magnitude larger than the measured vacuum energy density for the observable universe which is determined on the basis of the Standard Model of cosmology and empirical data. This enormous disparity between the expectations of two of our most widely accepted theoretical frameworks demands a credible and self-consistent explanation, and yet even after decades of sporadic effort, a generally accepted resolution of this crisis has not surfaced'.

In this book the subject of vacuum density remains open. Research of Kovalenko and others speculates the vacuum might be a four-dimensional reality. Article was published in Physics Letters B back in 2005 (Three-dimensional vacuum domains in four-dimensional SU(2) gluodynamics, Physics Letters B, **613**, 1–2, 52–56,): 'It is a general trend in modern theoretical physics to consider extended objects, like strings and membranes. Usually, one applies these ideas to hypothetical, high-dimensional completions of the four-dimensional world. However, lower-dimensional structures might also exist in four dimensions. At the present time, there is no well-developed theory which would predict such structures. However, there is accumulating evidence obtained within the lattice QCD that there are lower dimensions objects percolating through the vacuum of four-dimensional Yang–Mills theories'". If the vacuum actually is four dimensional, we cannot apply a classical understanding of vacuum density, which works only in the three-dimensional domain.

Rather, I will show the relatedness between the variable density of the vacuum and the variable rate of clocks. With a variable rate of clocks, we can indirectly measure the variable density of the vacuum. In General Relativity, the gravitational time dilation is calculated using the following formula:

$$t = \frac{t_0}{\sqrt{1 - \dfrac{2GM}{rc^2}}},$$

where t_0 is the rate of the clock on the surface of the stellar object, M is the mass of the stellar object, G is the gravitational constant, r is the radius of the stellar object and t is the rate of the clock at the point T which is infinitely away in empty cosmic space. For example, when one second has passed on the Earth surface, at the point T in infinity 1.000000000695915 second has passed. We can calculate the rate of a clock at point T_1, situated at the distance h above the surface of the stellar object with the following formula:

$$t = t_0 \cdot \sqrt{\frac{1 - \dfrac{2GM}{(r+h) \cdot c^2}}{1 - \dfrac{2GM}{rc^2}}}.$$

Let us calculate the time t at a point 20 km above the Earth's surface comparing with the 1 second elapsed time on the Earth's surface:

$$t = 1s \cdot \sqrt{\frac{1 - \dfrac{2(5.97219 \times 10^{24} kg)(6.67408 \times 10^{-11} m^3 kg^{-1} s^{-2})}{(6371000m + 20000m)(8.99 \times 10^{16} m^2 s^{-2})}}{1 - \dfrac{2(5.97219 \times 10^{24} kg)(6.67408 \times 10^{-11} m^3 kg^{-1} s^{-2})}{(6371000m)(8.99 \times 10^{16} m^2 s^{-2})}}}$$

$$t = 1s \cdot \sqrt{\frac{1 - 0.00000000138747}{1 - 0.00000000139183}}$$

$$t = 1.00000000000218s$$

(20 km above the surface)

Let us calculate the time t at the point 40 km above the Earth's surface compared with the 1 second elapsed time on the Earth's surface:

$$t = 1s \cdot \sqrt{\frac{1 - \dfrac{7.9717748 \times 10^{14} m^3 s^{-2}}{(6411000m)(8.99 \times 10^{16} m^2 s^{-2})}}{1 - \dfrac{7.9717748 \times 10^{14} m^3 s^{-2}}{(6371000m)(8.99 \times 10^{16} m^2 s^{-2})}}}$$

$$t = 1s \cdot \sqrt{\frac{1 - 0.00000000138315}{1 - 0.00000000139183}}$$

$$t = 1.00000000000434s \quad \text{(40 km above the surface)}$$

Let us calculate the time t at the black hole with the mass of the Sun and radius of 3000 metres compared with the elapsed $t_\infty = 1,000000000695915s$:

$$t_\infty = \frac{t_{black-hole}}{\sqrt{1 - \dfrac{2GM}{rc^2}}}$$

$$t_{black-hole} = 1.0000000006\ 95915\ s \cdot \sqrt{1 - \frac{2 \cdot 1{,}989 \cdot 10^{30} kg \cdot 6.67408 \cdot 10^{-11} m^2 kg^{-1} s^{-2}}{3 \cdot 10^3 \cdot 8.99 \cdot 10^{16} m^2 s^{-2}}}$$

$$t_{black-hole} = 1.000000000695915s \cdot \sqrt{1 - 0.98440824026696}$$

$$t_{black-hole} = 0.12486696822s$$

Black hole surface $t_{black-hole} = 0.12486696822s$

Earth surface $t_0 = 1s$

20 km above Earth surface $t_{20} = 1.00000000000218s$

40 km above Earth surface $t_{40} = 1.00000000000434s$

Infinite distance from Earth surface $t_\infty = 1.000000000695915s$

The rate of clocks is increasing with increasing vacuum density. Where the density of the vacuum is at the maximum ρ_{max}, the rate of clocks is at the maximum. With the diminishing of vacuum density, the rate of clocks is diminishing. The General Relativity effect causes clocks on the GPS satellites to run faster than on the Earth's surface by 45 microseconds per day. This is because on the satellite trajectory the vacuum is denser than on the Earth's surface.

GPS satellites are moving with a velocity v with respect to the Earth's surface. Because of its kinetic energy, the mass m of a given satellite is increasing:

$$m = m_0 + \frac{m_0 v^2}{2c^2}, \quad (14)$$

where m_0 is the mass of the satellite on the Earth's surface,

v is the velocity of the satellite relative to the Earth's surface. Because of the increased mass m of the moving satellite, the density of vacuum inside the satellite additionally decreases. The decrease of vacuum density causes clocks to run slower on the satellite than on the Earth's surface. The value of this Special Relativity effect is 7 microseconds per day.

The variable rate of clocks is directly related to the variable vacuum density. We could numerically evaluate the vacuum density on the surface of a given stellar object by considering that the numerical value of the vacuum infinitely distant from the stellar object is $\rho_\infty = 1.000000000695915$. On the Earth's surface the numerical value of vacuum density is $\rho_{earth} = 1$. On the black hole surface the numerical value of vacuum density $\rho_{black-hole} = 0.12486696822$.

In 20th century physics, the unsolved question was whether inertial mass and gravitational mass are caused by the mass of the given stellar object or are related to the masses of other stellar objects in the universe: 'If the rest of the universe determines the inertial frames, it follows that inertia is not an intrinsic property of matter, but arises as the result of matter with the rest matter of the universe. This immediately raises the problem of how Newton's laws of motion can be accurate despite their complete lack of reference to the physical properties of the universe, such as the amount of matter it contains'. The results of this research confirm that inertial mass and gravitational mass of a given stellar object with the mass m have their origin only in its mass, which causes the variable density of vacuum $\Delta\rho$, and are not related to the masses of other stellar objects.

The variable density of vacuum in proton and Higgs potential

In this chapter variable density of vacuum will be interpreted as the Higgs potential. Recent research of Melo I. published in the article »Higgs potential and fundamental physics« published in European Journal of Physics (*Eur. J. Phys.* **38**, 065404, 2017), presents the Higgs potential as follows: The Higgs potential $V(H)$ for a simple case of a real scalar field H can be written as:

$$V(H) = \lambda(H^2 - v^2)^2 = \lambda H^4 - 2\lambda v^2 H^2 + \lambda v^4$$

where H is the Higgs field. Both v and λ paramaters are determined experimentally through the measurement of the Fermi constant GF and the Higgs boson mass $MH = 125 GeV$, yielding $v = 246 GeV$, $\lambda = 0{,}13$. $V(H)$ can be interpreted as the Higgs vacuum energy density (energy density of the empty space). For our choice of the potential, the vacuum energy density is zero at the minimum $H = v$. However, for the potential energy it is the difference that matters, not the absolute value and thus the relevant contribution is the constant term in Eq. 24 (the size of the hill at $H = 0$), $\lambda v^4 = 4.8 \cdot 10^8 GeV^4$. From cosmology we have a vacuum energy density that is roughly 55 orders smaller and this huge difference is a mystery, the cosmological constant problem".

In the model here presented density of the vacuum in interstellar space has the value of ρ_{max}. We do not know yet the actual value of ρ_{max} which presents the actual cosmological constant problem as it is mentioned above by Melo.

5% of the energy in the universe is ordinary matter. The 65% percent is missing dark energy and 27% is missing dark matter. Considering that universal space has its physical origin in the vacuum, the energy of the vacuum itself can be the missing dark energy and the missing dark matter. The energy of the vacuum is not interacting with the light and remains invisible and undetectable.

The idea of 20th-century physics was that stellar objects exist in an empty space deprived of physical properties. This idea has led to the prediction of dark energy and dark matter. With the introduction of the vacuum which has variable density the question of dark energy and dark matter is seen from the new perspective which is promising to advance the solution for the cosmological constant problem. On the other hand, considering that vacuum could be a four-dimensional reality, the density of the vacuum could remain an open subject for a longer period of time because density (or energy density) is seen in today physics as three-dimensional phenomenon.

The idea that vacuum energy density is zero at the minimum $H = v$ is questionable. If we define vacuum energy density value is zero, then universal space could not exist anymore, because the vacuum is the physical origin of the universal space. The vacuum is the physical origin of the universal space; vacuum energy density (or density in the model presented in this article) is variable and bigger than zero in entire universal space.

The model presented in this article suggests that minimal vacuum density ρ_{min} on the surface of the proton placed in interstellar space is at the bottom of the hat, the density of vacuum in the centre of the proton ρ_{centre} is on the top of the hat, the density of the vacuum away from the proton ρ_{max} is on the edge of the hat.

In the proton, vacuum fluctuations are moving from the ρ_{max} to the ρ_{min}, and from the ρ_{centre} to the ρ_{min}. These vacuum fluctuations are the physical origin of the Higgs potential.

The superfluid quantum vacuum model with the variable density is the development of the electromagnetic quantum vacuum model (QED) which is one of the most successful theories. By giving electromagnetic vacuum variable density as presented in this article, we can describe Higgs potential and also the origin of gravity. The perspective of further research on the variable density of vacuum is to integrate QED with the Higgs mechanism model and quantum gravity model.

Recent research of Sbitnev on the hydrodynamics of the physical vacuum opens the new perspective where elementary subatomic particles could be seen as the vacuum vortexes. In Sbitnev model the vortex is periodically exchanging energy with the vacuum via vacuum fluctuations. Sbitnev model is enhancing the model of vacuum fluctuations presented in this article with clear insight, namely, we cannot study subatomic particles without considering their active relatedness with the vacuum.

According to the model presented in this book, a given vortex is in active relation with the vacuum. When accelerated the vortex is "dragging" with the vacuum and absorbing some of its energy which is its relativistic energy.

Considering that vacuum is 4-dimensional and so proton is 4-dimensional vacuum vortex, we are limited in the proton observation with the 3-dimensional apparatuses and 3-dimensional sensorial sense (sight). Taking into account that atom is three dimensional, the subatomic world could be four and more dimensional. We have to be aware that higher dimensionality of the subatomic world represents the limitation of our scientific endeavor.

When modelling mass–energy equivalence, inertia and gravity, we cannot develop an objective model without considering that space has physical properties. With the introduction of the superfluid quantum vacuum, which is the physical origin of the universal space, the new perspective presented in this article is open. This model confirms that inertial mass and gravitational mass are equal and both have their origin in the vacuum fluctuations caused by the variable density of vacuum.

Dynamic Vacuum Relativity and GPS

The Global Positioning System (GPS) is a satellite system that provides accurate position, velocity, geolocation and time information to a GPS receiver that may be located in earth, sea and air. A GPS satellite functions by transmitting signals to equipment on the ground and the GPS receiver passively satellite signals. GPS has many applications in Navigation, astronomy, cartography, mapping, cellular telephony, disaster relief, radio occultation, clock synchronization, geotagging, geofencing, fleet tracking, air tracking, mining, tours, recreation, robotics, surveying, sports, tectonics, telematics and other uses.

The GPS system proves without any doubt that the relative rate of clocks on satellites relative to the Earth's surface is valid for all observers: in aeroplanes, trains, ships, cars. This experimental fact and everyday living reality suggest a revision of our understanding of the famous *gedanken* experiment with the observer on the train station and observer on the passing train. In the physics textbooks, we read that a clock on the station is running faster for the observer on the train. The clock on the train is running slower for the observer on the station. Both observers have their 'internal time' inside the reference system in which they are and 'external time' which is inside the other reference system. In this interpretation, we have four different times: proper time of the observer on the station, proper time of the observer in the train, external time of the observer on the station and external time of the observer in the train. GPS proves that relative velocities of clocks on the station and in the train are equally related to the rate of clocks on the satellites and are valid for both observers. If this were not so, GPS could not work properly.

The aim of this chapter is to apply the bijective research methodology in the interpretation of topics such as variable density of vacuum, Lorentz factor, planets precession, Sagnac effect, and gravity impact on light. We found that the Lorentz factor comes from variable density of vacuum and that planets precession and Sagnac effect is a consequence of the vacuum rotation.

Lorentz factor and variable density of the vacuum

GPS suggests the abolition of inertial systems in Special

Relativity (SR) and a search for a more elegant model which will better correspond to the physical world. In this improved model, the Lorentz factor γ expresses a diminished rate of clocks and diminished velocity of material changes in general and is defined as:

$$\gamma = \frac{1}{\sqrt{1 - \dfrac{v^2}{c^2}}}$$

where v is the relative velocity between inertial reference frames and c is the speed of light in a vacuum.

In a famous example with the train passing the station t' is elapsed time on the train and t is elapsed time on the station:

$$t' = \gamma \cdot t$$

This diminished rate of clocks on the train has its origin in the decreased density of vacuum inside the train. In general a moving system is interacting with the vacuum energy; the higher the velocity v, the stronger the interaction and more vacuum energy is integrated into the moving object which increases its mass m:

$$m = \frac{1}{\sqrt{1 - \dfrac{v^2}{c^2}}} \cdot m_0$$

where m_0 is object's rest mass.

Mass m_0 of the given physical object at rest can be expressed with the diminished energy of the vacuum on its surface as it follows:

$$m_0 = (\rho_{max} - \rho_{min}) \cdot V$$

where m_0 is the mass of the object at rest , ρ_{max} is density of the vacuum in interstellar space, ρ_{min} is density of the vacuum on the surface of the physical object, V is the volume of the object.

For the moving physical object mass m can be expressed as follows:

$$m = \gamma \cdot m_0 = (\rho_{max} - \rho_{minR}) \cdot V$$

where γ is Lorentz factor, m_0 is rest mass, and ρ_{minR} is density of the vacuum on the surface of the physical object. Mass of the moving physical object is increasing because moving object is interacting with the vacuum and absorbing some of its energy. Equation confirms that because of its speed, v, the train's mass increases and the density of vacuum ρ_{minR} on the moving train surface decreases additionally regarding the train at rest. This decreased density of vacuum causes the rate of the clock in the train to run slower. According to equation above, we can write the following formula for the Lorentz factor:

$$\gamma = \frac{(\rho_{max} - \rho_{minR}) \cdot V}{m_0}$$

which confirms that the Lorentz factor depends on the density of vacuum in the surface of the moving object ρ_{minR} , of the density of the vacuum in interstellar space ρ_{max} and the volume, V, of the object. The minimal density of vacuum ρ_{minR} depends on the speed v of the physical object. The higher is the speed v, stronger is interaction of the object with the vacuum, absorption of the vacuum energy is bigger and density of vacuum on the surface of the object ρ_{min} is getting smaller. Smaller is density of the vacuum slower is the rate of the clock on the surface of the physical object:

increased velocity $\rightarrow$ absorption of the vacuum energy is increasing $\rightarrow$ density of the vacuum is decreasing $\rightarrow$ rate of clocks is decreasing

For smaller physical objects (compared with stellar objects), for example, satellites, trains, minimal density on their surface is a good approximation for the density of the vacuum inside the object. This is because the difference between the density of the vacuum on the surface and the density of the vacuum in the center of the object is minimal. This is not valid for stellar objects where the density of vacuum increases by Newton's shell theorem when going to the centre of the stellar object.

In general, we can conclude that the relative rate of clocks depends only on the variable density of the vacuum. For example, muons' decay when approaching the Earth's surface decreases because their velocity increases. Because of their increasing velocity, the minimal density of vacuum on their surface decreases, and time of decay increases. The duration of this decay does not depend on the chosen reference system and chosen observer; muons' relativistic decay is valid for all observers and is determined only by the variable density of the vacuum.

Special Relativity Theory in Dynamic Vacuum

The Lorentz factor is not related to some inertial system; it depends only on the variable density of the vacuum. In this perspective, the vacuum is the absolute reference system in which elementary particles, physical objects and stellar objects move. Einstein's idea that there is no absolute motion is examined here again. Because of the abolishment of ether at the end of the 19th century, the idea of absolute motion was also abolished. Introduction of vacuum as the fundamental arena of the universe allows the reintroduction of absolute motion, which is the motion of a given physical object in a vacuum as the absolute reference frame.

With the abolishment of the ether, Einstein introduced the 'relative motion' concept, which means that there is no absolute reference system in which a given motion occurs.

According to Einstein, a given moving object's velocity can be measured only in comparison with another object's velocity. I will show now that this Einstein idea makes no sense.

It has been already shown in this book that light speed is constant for all observers because all observers exist in the same medium called the vacuum and light is the vibration of the vacuum. The vacuum is the physical origin of universal space which is not 'empty'; we can call universal space 'physical space'.

In physical space which is the absolute frame of reference, we distinguish relative and absolute velocity. You imagine you drive a car with a speed of 110 km per hour. 500 meters in front of you is a car moving with a speed of 120 km per hour. An accident happens and the car in front of you crashes in the wall along the street. The driver luckily survives but the car is totally destroyed. If you measure the velocity of the car in front of you from your car you will measure 10 km per hour. This simple example proves that the "relative speed" of 10 km per hour of the car in front of you was not "real". The car would not be totally destroyed if it would crash with the wall with 10 km per hour. The real velocity of the car is the velocity regarding the highway, which is 120 km per hour.

The same is valid for the velocity of the light, whose "real" velocity is its velocity in the physical space and is valid for all observers.

By assuming the vacuum as the absolute reference frame, we can develop SR theory without contradictions. The existing SR has one contradiction in the thought experiment of two photon clocks. We place two identical photon clocks on the moving train, one is positioned horizontally in the direction of motion; the other is positioned vertically. According to the 'length contraction', the horizontal photon clock will shrink in length, the vertical will not. The horizontal photon clock will shorten and will tick faster than the vertical photon clock which will not diminish in length. This leads to a contradiction, SR does not predict that two clocks in the same inertial system have different rates. The solution is the development of SR in three-dimensional Euclidean space with Galilean transformation and Selleri formula for the variable rate of clocks. SR equipped with such formalism describes successfully all phenomena which are described by classical SR.

The second contradiction is the rate of the vertical photon clock on the moving train for the observer on the station. The physics textbooks say that for the observer on the station, the vertical photon clock is ticking slower because he sees that the photon is moving in a 'zig-zag' direction:

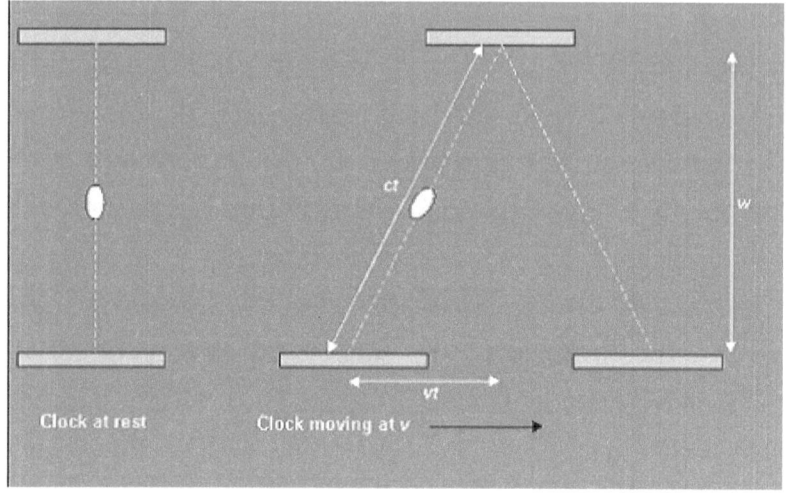

The observer at rest seeing the moving clock photon

This seems an awful explanation. The optical illusion of the stationary observer cannot slow the rate of the clock. The fact is that in a moving train the density of vacuum diminished and because of this the velocity of the photon diminishes too; the moving photon clock is actually ticking slower and its rate is valid for all observers.

Back in 1964, Shapiro measured to be the decreased velocity of light in a gravitational field. The speed of a light signal diminishes when passing the gravitational field of the Sun. Shapiro's result is understood in today's physics as 'gravitational time delay' which is caused by spacetime dilation, which increases the path length. According to bijective research methodology, this interpretation seems not exact. Shapiro did not measure space–time dilation. In the

Bijective physics, the element of 'spacetime dilation' has no bijective correspondence in the physical world. Never has it been observed in physics that spacetime or space are dilating. According to bijective research methodology, Shapiro's result should be named 'gravitational diminishing of light speed' which is caused by the diminished density of the vacuum. In the area of universal space with a given gravitational field, the density of the vacuum is diminished. This diminished density of the vacuum causes minimal diminishing of light speed which is defined by the permittivity and permeability of the vacuum:

$$c = \frac{1}{\sqrt{\mu_0 \varepsilon_0}} \, ,$$

where μ_0 is the magnetic permeability of vacuum and ε_0 is electric permittivity of the vacuum where there is no influence of gravity and density of vacuum is at maximum ρ_{max}. In the space with a gravity field the density of vacuum decreases. This decreased density causes minimal diminishing of permittivity and permeability which causes minimal diminishing of the light speed.

Masanory says in his article published on arXiv (Masanori S., Gravitational effect on the refractive index: A hypothesis that the permittivity, εo, and permeability, μo are dragged and modified by the gravity): "It is known that the speed of light depends on the gravitational potential. In the gravitational fields, the speed of light becomes slow and time dilation occurs".

In Dynamic Vacuum Relativity, the permittivity and permeability of free space are assumed to depend on gravity and are variable. Minimal variability of light speed caused by gravity field keeps SR in place because the first postulate of SR is valid only in space with the absence of gravity.

It will be shown in this chapter that the Doppler effect proves the second postulate of SR. The second postulate of SR is: 'The speed of light c is constant and is independent of the relative motion of the source'. In this article it is shown that the observer exists in the vacuum and a photon is the vibration of the same vacuum. When the observer is moving towards the source of light or away from the source, he will experience the Doppler effect. With the understanding that the moving observer and the source both exist in the same medium (vacuum) and that light is the vibration of the vacuum, the second postulate becomes logical. You observe the light coming from the source. Light has a given frequency. When you start moving away from the source, or closer to the source, the frequency of light diminishes or increases, respectively. In the Doppler effect, the velocity of light c for the moving observer remains constant because light is the wave of the medium in which the observer moves.

General Relativity Theory in Dynamic Vacuum Model

In General Theory of Relativity, a given physical object is increasing the curvature of space. In Dynamic Vacuum

Relativity is developed a similar model where a given physical object is diminishing density of the vacuum. Increasing of mass in the given area of space decreases the density of the vacuum. Bigger is the curvature of space, smaller is the vacuum density. The rotation of stellar objects also causes rotation of the vacuum. Rotation of the vacuum around the Sun causes precession of the planets according to the following equation:

$$\sigma = \frac{24\pi^3 L^2}{Tc^2(1-e^2)},$$

where perihelion shift σ is expressed in radians per revolution, L is the semi-major axis, T is the orbital period, c is the speed of light, and e is the orbital eccentricity. We can see in this formula there is no mass of the Sun, there is no mass of a given planet which means that masses of the Sun and planets do not affect the precession of the planets directly. In the model of dynamic quantum vacuum the perihelion shift σ depends on the rotation of the vacuum which is caused by the rotation of the Sun. The rotating vacuum is pushing the planets and causing their perihelion precession. On the basis of this model, the mathematical formalism was developed. The calculations confirm the same results for the precession of all planets as was calculated by Einstein.

In a GPS system, Sagnac effect corrections make the whole system work. The article on this intertsting subject was published by Ashby N. (The Sagnac Effect in the Global Positioning System. In: Rizzi G.,   Ruggiero M.L. (eds) Relativity in Rotating Frames. Fundamental Theories of Physics, vol 135. Springer, Dordrecht, 2004). Essentially, what is corrected is that the signal when moving from A to B in the direction of Earth's rotation needs less time than when moving in the direction from B to A, which is opposite to Earth's rotation. In a vacuum, light has a constant speed regardless of the vacuum motion. When moving from A to B, light needs less time because it is moving in the same direction as the vacuum.

The Michelson–Morley experiment has givren a null result because the area of vacuum around the Earth is not only rotating with the Earth, it is also moving with the Earth. The negative outcome of the Michelson–Morley experiment abolished the ether model. According to bijective research methodology, universal space is neither filled with the ether nor it is 'empty'. Universal space contains material objects which have their energy. According to bijective methodology, energy and matter cannot exist in an 'empty' space deprived of all physical properties. Universal space has its physical origin in the vacuum. A photon is a vibration of the vacuum. The velocity of this vibration expansion is the speed of light, c. Photon velocity c is invariant regarding the vacuum motion as it appears in the Sagnac effect. Photon velocity diminishes minimally when a photon is moving in the stronger gravity where the density of vacuum is lower, as is the case with the Shapiro experiment.

Motion and rotation of the vacuum with physical objects is called in this book the 'vacuum dragging effect'. This effect was measured by Josef Lense and Hans Thirring back in 1918 and was called 'Frame-dragging' as an effect on spacetime being distorted by rotating objects (Lense, J., Thirring H., "Über den Einfluss der Eigenrotation der Zentralkörper auf die Bewegung der Planeten und Monde nach der Einsteinschen Gravitationstheorie". Physikalische Zeitschrift. **19**: 156–163. [On the Influence of the Proper Rotation of Central Bodies on the Motions of Planets and Moons According to Einstein's Theory of Gravitation], *1918*).

Bijective physics confirms that the 'space–time' model has no physical reality and so it cannot be dragged by rotating or moving objects. According to bijective research methodology, a more adequate term is 'vacuum dragging effect'.

Dynamic Vacuum Impact on Light

Gravitational redshift

Gravity has its origin in vacuum fluctuations which are interacting with photons and diminishing their frequency; we call this 'gravitational redshift'. When light from distant galaxies reaches the Earth, its frequency becomes lower. On the path to the Earth, light loses some of its energy because it is moving against the gravity flow of the vacuum, which is pointing into the direction of galaxies:

$$E_{photon.Earth} = E_{photon.galaxy} - \Delta E \, ,$$

where $E_{photon.galaxy}$ is the energy of the photon on the galaxy, $E_{photon.Earth}$ is the energy of the arrived photon on the Earth, and ΔE is the loss of the energy due to the gravity flow of the vacuum.

$$\Delta E = h \cdot \Delta v \, ,$$

where h is the Planck constant and Δv is the decrease of photon frequency because of gravitational vacuum fluctuations.

Because of different density of the vacuum and correspondent vacuum fluctuations which carry gravity, frequency of light is also changing when moving from the source to the receiver on the earth surface. In Harvard university experiment was done where the source was on the earth surface and the receiver was on the height of 22,5 meters.

It was accomplished in this experiment that by the use of Mössbauer-effect measurement of the difference between y-ray emission and absorption frequencies between the source on the Earth surface and receiver 22 m above. The measurement accuracy was $\Delta\omega / \omega \approx 10^{-15}$, the formalism which shows change of light frequency is following :

$$\frac{\Delta\omega}{\omega} = \frac{GM}{R^2 c^2} \cdot h$$

where M and R are the mass and radius of the Earth. The source of data presented above is the article of American physicist Puthoff H.E. (Polarizable-Vacuum (PV) presentation of general relativity, Found.Phys. 32, 927-943. 2002).

We can substitute in formula mass M with the $(\rho_{max} - \rho_{min}) \cdot V$, and we can write the following formula:

$$\frac{\Delta\omega}{\omega} = \frac{G \cdot (\rho_{max} - \rho_{min}) \cdot V}{R^2 c^2} \cdot h$$

Formula (14) can be developed as follow:

$$\frac{\Delta\omega}{\omega} = \frac{G \cdot (\rho_{max} - \rho_{min}) \cdot 4\pi R^3}{3R^2 c^2} \cdot h$$

$$\frac{\Delta\omega}{\omega} = \frac{4\pi R \cdot G \cdot (\rho_{max} - \rho_{min})}{3c^2} \cdot h$$

Formula confirms that gravitational red shift depends on the minimal density of the vacuum ρ_{min} on the earth surface. Minimal density of the vacuum on the earth surface is the origin of the vacuum fluctuations which are fluctuating from the outer space towards the earth.

Shapiro gravitational delay

Electric permittivity in flat space with no gravity is ε_0 Magnetic permeability in flat with no gravity space is μ_0 Following Puthoff [14] on the surface of stellar object permittivity and permeability get following values:

$$\varepsilon = K \cdot \varepsilon_0$$

$$\mu = K \cdot \mu_0$$

where vacuum dielectric constant K on the surface of stellar object is following:

$$K \approx 1 + \frac{2GM}{rc^2}$$

In equation above G is the gravitational constant, M is the mass and r is the distance from the origin located at the center of the mass M.

I introduce here heuristic equation:

$$K = \frac{\mu}{\mu_0} = \frac{\varepsilon}{\varepsilon_0} = \frac{\rho_{max}}{\rho_{min}}$$

Equation above shows that the vacuum dielectric constant depends on the variable density of the vacuum. In this sense diminished density of the vacuum is increasing permittivity and permeability of the vacuum which decreases minimally velocity of light as follows:

$$c = \frac{1}{\sqrt{\varepsilon \cdot \mu}}$$
,

where ε is permittivity of the vacuum in the gravitational field and μ is permeability of the vacuum in gravity field. Out of this follows that Shapiro gravitational time dilation has the origin in the diminished density of the vacuum near the stellar objects which increases dielectric constant K of the vacuum and minimally decreases the velocity of light.

diminishing of vacuum density $\Rightarrow$ increases dielectric constant $\Rightarrow$ increases electric permittivity of the vacuum $\Rightarrow$ increases magnetic permeability of the vacuum $\Rightarrow$ decreases velocity of light

The official texts book explanation of the Shapiro experiment is that in stronger gravity time as the 4th physical dimension of the space is dilating and so light needs more time to reach from the point A to the point B in the space-time which is the fundamental arena of the universe. I have shown that bijective interpretation of data (where data are not interpreted but read directly) requires exact explanation, namely, the velocity of light is minimally diminishing in gravity field due to the diminished density of the vacuum.

Gravitational lens

Vacuum fluctuations are bending the light, which we call 'gravitational lens'. The bending of light when passing the Sun is one of the proofs of General Relativity. What actually happens is that vacuum fluctuations near the Sun's surface push the photons and this pushing causes the bending of the light Gravity vacuum fluctuations are bending the photon's trajectory, we call this 'gravitational lens'. The Einstein formula for the bending of light when passing the Sun is the following:

$$\delta = \frac{4G \cdot M_S}{c^2 \cdot b},$$

where δ is the angle of deflection, M_S is the mass of the Sun, c is light speed, and b is the minimum distance between the trajectory and the centre of the Sun. The mass of the Sun, M_S, can be expressed according to formula:

$$\delta = \frac{4G \cdot (\rho_{max} - \rho_{min}) \cdot V}{c^2 \cdot b}.$$

Formula above confirms that vacuum fluctuations which carry gravity are directed from the outer space where density of the vacuum is ρ_{max} towards the surface of the Sun, where density of the vacuum is ρ_{min}, see Figure 5. These vacuum fluctuations push the photons and cause light deflection. Bijective research methodology confirms that light when passing the Sun is not deflected as a result of space being curved. NASA measurements have proved universal space has a Euclidean shape. Curvature of space in GR is only the mathematical description of the variable density of the vacuum. Actual universal space has Euclidean shape. Light is deflected because vacuum gravitational fluctuations push light when passing the Sun.

Technological applications of dynamic vacuum model

The novelty of the model presented in this book is the variable density of the vacuum. By developing technology which will be able to manipulate density of the vacuum we will develop antigravity. Imagine a spaceship on the earth surface which can increase vacuum density. With increasing of the density of the vacuum, the spaceship will move in the direction of maximum vacuum density which means moving away from the earth surface towards interstellar space. Density of the vacuum on the earth surface is $\rho_{min} = \rho_{max} - 5.514 \cdot 10^3 \, kg/m^3$. With the local increase of the vacuum density ρ_s inside the spaceship the area of local increase will move in the direction of maximum vacuum density ρ_{max}. Spaceship will stop when its local vacuum density ρ_s will be equal to the vacuum density outside the spaceship.

The development of Relativity Theory as we have seen in this chapter led to the abolishment of space-time as the fundamental region of the universe and its replacement with the vacuum. The vacuum is dynamic in the sense that it rotates and moves with the physical objects. Light is the vibration of the vacuum; its speed is invariant from the rotating and moving vacuum. All 'relativistic phenomena' have their origin in the variable density of the vacuum and in the 'vacuum dragging effect'.

6
Bio-Universe Model

Cosmology, in order to give good results, has to start with the shape of the universal space. NASA result is confirming with the NASA measurements have proved universal space has a Euclidean shape with only a 0.4% margin of error. You type in Google "Will the universe expand forever?" + NASA and article will pop out.

This means that the right model in cosmology is Euclidean space. We cannot use Riemann finite geometry in the description of the universal space. NASA results confirm universal space is infinite, which means the amount of the energy in the universe is infinite. The universe cannot be approached as the finite system, the universe is an infinite system.

We can define the position of the object A in universal space regarding object B with three coordinates. This does not mean that universal space is 3-dimensional. About the dimensionality of the universal space, we do not know yet and probably we will never know. As mathematics allows that Euclidean space has more than three dimensions, we cannot exclude, that universal space is multidimensional too. Article of Bronnikov and others titled "Inhomogeneous compact extra dimensions" (arXiv 2017) suggest that the universe has inhomogeneous compact extra dimensions. In this article is developed the idea that these extra dimensions have a physical basis in the dynamic vacuum which is multidimensional. Gravity is the result of vacuum fluctuations of the 4ᵗʰ dimension of the multidimensional dynamic vacuum (vacuum). A 3-dimensional physical object is diminishing the density of the 4ᵗʰ dimension of dynamic vacuum exactly for the amount of its energy:

$$\frac{E}{c^2} = m = (\rho_{max} - \rho_{min}) \cdot V \quad ,$$

where ρ_{max} density of the vacuum in the interstellar space, ρ_{min} is density of the vacuum on the surface of the physical object, V is the volume of the object. Vacuum fluctuations which flow from ρ_{max} towards ρ_{min} are the origin of gravity which is the main force of universal dynamics.

Higher Dimensions of the Vacuum in the living Organism

In Bijective physics vacuum model is replacing space-time as the fundamental arena of the universe. In vacuum time has merely mathematical existence, namely, time is the numerical sequential order of events running in a vacuum. The vacuum is timeless in the sense time is not its 4th dimension. Vacuum is the direct information medium of entanglement regarding EPR-type experiments. This function of the dynamic vacuum can also be extended on living matter, means living organisms, where there is a permanent communication between cells via coherent electromagnetic fields which can be seen according to quantum electrodynamics (QED) as the formations of the vacuum itself. This experimental data opens new perspectives in biology, where a living organism is seen as living physical object in which vacuum plays an important communication role in intercellular communication.

The idea of relatedness of life with vacuum was also studied by Hungarian musician and physicist Grandpierre Attila and Slovene biologist Igor Jerman. In the perspective of presented researches above dynamic vacuum is governing the active behavior of living organisms via the electromagnetic field. In the living organism, we have electromagnetic fields which are interrelated with atomic and molecular structures of the living organism.

Higher-dimensional vacuum fluctuations in a living organism are stronger than in the same dead organism. We can express this energy difference between living and the same dead organism as follows:

$$E_{life} = E_{dead} + E_v \, ,$$

where E_{life} is energy of living organism, E_{dead} is energy of dead organism, and E_v is higher dimensional vacuum fluctuations which are characteristic for the living organism. This is the working hypothesis of Bio-Universe model.

Slawinski has measured, that vacuum fluctuations in form of electromagnetism vanish at the time of death: biophoton emission at the time of death is about 10 to 100 time stronger than biophoton emission of the living organism. Biophoton emission cannot influence the mass of the dead organism because the emission of the biophotons does not change the nuclei of the atoms which build living organism. The mass of a given system is decreasing when nuclei are falling apart. This happens for example in an atom bomb and a huge amount of electromagnetic energy in the form of heat and light is released. Despite biophoton emission, we can write the following formula:

$$m_{life} = m_{dead} \; ,$$

where m_{life} is mass of the living organism, m_{dead} is the mass of dead organism.

Equality of living mass and dead mass is in accordance with today physics. You imagine you frozen a living organism in a time of a microsecond. The atoms before frosting and after frosting are the same, so the mass is the same. However, experimental data obtained back in 1987-90 at the University of Ljubljana shows that gravity is stronger on the living organism as on the same dead organism. Article with these results I published with my research fellows back in 2017 in NeuroQuantology. The article is titled "Unified Field Theory Based on Bijective Methodology".

Material and Methods

By earthworms (lumbricus teresticus) the measured difference of gravity on the alive and the same dead organism is about one million part of living mass. 5 grams of earthworms lost at the time of death about 5 micrograms, 70 grams of the earthworms lost about 70 micrograms, which we express with the formula below:

$$F_{g(life)} = F_{g(dead)} + \Delta F_g .$$

The experiment was repeated 12 times. Statistical analysis of the obtained data confirms that data are right within 95%. In order to exclude human factor, the experiment was repeated six times in April 2019 on highly precise Mettler-Toledo mass comparator AX107H where data are directly transferred from the mass comparator to the computer without human reading. The same results as back in 1987-90 were obtained. Weight of the 5 grams of living worms diminishes 15 minutes after dead for 5 micrograms. The weight difference remains the same for the next two hours. The average loss of the weight by earthworms is 1 million part of the living weight.

Repeating the experiment with the worms is the necessary killing of the living organisms in order to prove the existence of the higher dimensional layers of the vacuum on the gravity. The author does not encourage researchers to repeat this experiment with the higher develop organisms.

The diminished gravity force on a dead organism cannot be explained in the frame of Einstein's "mass-energy equivalence", namely, that bio-photon emission at death could cause such a diminishing of energy which would result in diminishing of gravity for ΔF_g. When 10 micrograms of matter is transformed in electromagnetic energy (according to the formula $m = E/c^2$) a huge release of energy is expected. There is no such a release of energy at the time of death. The "gravity gap" ΔF_g between living and the same dead organism in this article is interpreted as a result of higher-dimensional vacuum fluctuations in the living organism which are characteristic for living organism.

Weight of the living organism has two components. First is mass m of the living organism, which diminishes the density of the vacuum and generates vacuum fluctuations which carry gravity. The second component of gravity is higher dimensional vacuum fluctuations which are characteristic for living organism.

Mettler-Toledo mass comparator AX107H
ith 2 test tubes with distilled water
and two test tubes with worms

Higher dimensional vacuum fluctuations and "Orch OR" theory

The experiment with the worms proves that higher-dimensional quantum fluctuations increase gravity force on the living organism. This is in favor of Orch OR theory which sees life and consciousness as phenomena which are deeply related to the structures of the universe: "The DP form of OR is related to the fundamentals of quantum mechanics and space-time geometry, so Orch OR suggests that there is a connection between the brain's biomolecular processes and the basic structure of the universe" (Stuart Hameroff, Roger Penrose, Consciousness in the universe A review of the 'Orch OR' theory, Physics of Life Reviews 11 (2014) 39–78).

In **Bio-Universe Model** (from now on BUM) multidimensional vacuum with n dimensions (n is the cardinal number of natural numbers) is primordial structure of the universe which governs the evolution of life on the planet Earth and in the entire universe. In BUM consciousness is mathematically described as the energy of the photon in an n-dimensional layer of vacuum which has infinite frequency:

$$E = n \cdot h \cdot v ,$$

where v represents frequency of the photon, h is Planck constant, integer n represents the infinite dimension of the vacuum. Consciousness is the vibration of the n-dimensional layer of the vacuum where the frequency is infinite value the wavelength value is zero. Out of that, it follows that the velocity v of consciousness is zero:

$$v = \infty \cdot 0 = 0 .$$

Consciousness exists in the entire universe; it is the most fundamental property of the universe. Entanglement of EPR type phenomena is transferred via consciousness as the n-dimension of the vacuum. The mathematical model of consciousness cannot be considered as a bijective model of consciousness; it only points towards its real nature. Consciousness is subjective and should be searched in a subjective way by watching the way scientific mind functions. The more one is developing the ability to observe, to watch his mind, the more one enters the observer, which is consciousness itself.

Consciousness is manifesting in the rational part of the human mind as mathematics. The universe is ruled by mathematical laws. The Founder of "Mathematical Universe Hypothesis" (MUH), Max Tegmark, says: Our external physical reality is a mathematical structure" (Tegmark M., The Mathematical Universe, Foundations of Physics. **38** (2): 101–150. 2008).

Tegmark statement is close to the truth because we cannot approach matter apart from the multidimensional vacuum. Valeriy Sbitnev has developed the model of vorticity dynamics, published on arXiv (Hydrodynamics of the Physical Vacuum, Vorticity dynamics, 2016). After reading his articles I think elementary particle, proton, electron, and photon can be seen as different vortexes of multidimensional vacuum. The mathematics exists in higher dimensional layers of the vacuum and is governing 3-dimensional physical world via 4 and 5-dimensional layers. In BUM atom is three dimensional, elementary particles as proton, electron, and photon are different vortexes of the 4-dimensional layer of the vacuum. What in Chinese medicine is called "QI energy" (life energy), in Indian medicine "Prana", in BUM is described by the 5-dimensional layer of the vacuum. The human mind is 6 and a more dimensional layer of the vacuum. That's why the human mind was able to develop the mathematics of n-dimensional spaces. If mind would have physical origin in the molecular structure of the brain which is 3-dimensional it could develop only 3-dimensional Euclidean geometry. All that we can develop in mathematics has an independent existence in the universe. Mathematics is the backbone of the universe.

3-dimensional atomic and molecular layer of life is the last extension of the multidimensional vacuum. Science urgently needs to extend research on the higher dimensional layers of the vacuum in order to enter into deeper synchronicity with nature and the universe. Human society is the small sub-system of the universe which is a far bigger and complex system. Without tuning with the universe we have low chances to survive as the species on this planet. Research on higher dimensional layers of the vacuum is subjective. In subjective research counts experience which does not need to be measured in order to be recognized as real. Measurement has limits already on the 4th layer of reality. We cannot fully grasp elementary particles which are 4-dimensional structures of the multidimensional vacuum because we observe them with three-dimensional apparatuses.

Beyond geocentric paradigm of life

BUM is surpassing geo-centrism of 20th-century biology which sees the evolution of life as phenomena which are evolving on the planet Earth surface. BUM is approaching life as a phenomenon occurring in the multidimensional vacuum which is the fundamental arena of the universe.

Experimental data presented in this article confirms that in living matter multidimensional vacuum is more active than in ordinary nonliving matter. This is valid also for gravity fluctuations which are stronger in living mass than in the same dead mass which results that the weight of the living organism is a bit bigger (by worms for about one million part) than the weight of the same dead organism. This experiment proves that life is intrinsically related to a dynamic vacuum which is the fundamental physical property of the universe.

Further on, more than 50 stable organic molecules have been detected in the interstellar medium. They are many planets with similar characteristics as our planet: "There is a large parameter space of possible physical characteristics of Earth-like extrasolar planets, and a more careful study of time variation and surface features is recommended". Thinking that life has evolved only on planet Earth has become unscientific. We have enough experimental data for a paradigm shift which reaches beyond geo-centrism. Considering vacuum as the fundamental arena of the universe we can take assume that vacuum represents the fundamental physical parameter which determines the development of life. With this view, we are surpassing the geocentric view on the evolution of life which is now seen as the larger universal process running in the entire universe. This book is a humble attempt to relate life and multidimensional vacuum, to connect biology and cosmology.

Right Interpretation of Data means the End of Big-Bang Model

In BUM time has only the mathematical existence. We do not have any experimental data, which would confirm that CMBR exists in some fundamental physical arena called "space-time". CMBR is measured only in universal space. This means that CMBR cannot have the origin in some remote physical past, because in the universe there is no past at all. Universal dynamics run in a dynamic vacuum where time has only the mathematical existence. Events in the universe are irreversible. We experience events that happened as "past", which is fine, but we have to understand that these events have no physical existence anymore and cannot be the source of radiation as for example CMBR. The only possible source of CMBR is the dynamic vacuum.

Cosmological redshift data are not proving the expansion of the universe. This is the wrong interpretation. When light from the distant galaxies is reaching the Earth, its frequency is getting lower. On the path to the Earth, light is losing some of its energy because it is moving against the gravity flow of the vacuum, which is pointing into the direction of galaxies:

$$E_{photon.Earth} = E_{photon.galaxy} - \Delta E \; ,$$

where $E_{photon.galaxy}$ is the energy of the photon on the galaxy, $E_{photon.Earth}$ is energy of the arrived photon on the Earth and ΔE is a loos of the energy due to the gravity flow of the vacuum.

$$\Delta E = h \cdot \Delta v \; ,$$

where h is Planck constant, Δv is diminishing of photon frequency because of gravitational vacuum fluctuations.

American astronomer of Swiss origin Fritz Zwicky has named this effect "tired light". He published an article on the subject back in 1929 titled "On the Red Shift of Spectral Lines through Interstellar Space«. My research is fully supporting his results.

Big Bang model was developed and popularized by Belgian catholic priest Georges Lemaitre which was in favor of the model which needs the creator. Big Bang model has religious elements which are rationalized by the "inflation theory" which is against the first law of thermodynamics. Big Bang model is the biggest failure of 20-century science. The Bio-universe model works without God, in this model Universe is God.

Recent research by Lawrence H. and others confirms that the idea of accelerating universe is not in accord with astronomical observations: "Using the largest available supernova data set, the JLA catalogue, we find that the timescape model fits the luminosity distance-redshift data with a likelihood that is statistically indistinguishable from the standard spatially flat Λ cold dark matter cosmology by Bayesian comparison. In the timescape case cosmic acceleration is non-zero but has a marginal amplitude, with best-fitting apparent deceleration parameter, $q_0 = -0.043 + 0.004 - 0.000$" (Apparent cosmic acceleration from type Ia supernovae, arXiv 2017).

Several data obtained in astronomy at the end of the last century have shown that observable universe is much larger than the universe according to the Big-Bang model which has the beginning and somehow needs the "creator". Thinking of the universe as something that could have the beginning is not scientific, rather is a religious view in cosmology. Big Bang model belongs to the history of physics. Universe is a non-created system in a permanent dynamic equilibrium; in black holes, the matter is continuously disintegrating back into the energy of the vacuum, in intergalactic space protons are continuously formed out of the dynamic vacuum energy. Albert Einstein used to say: "My scientific work is driven by an irresistible desire to understand the secrets of nature, and by nothing else". In order to progress cosmology, we need to start directly reading obtained data and drop creating models where these data are interpreted by our mind as it is the case with the Big-Bang model.

Evolution of life is a consistent part of universal dynamics. The universe is a non-created system in a permanent dynamic equilibrium. The idea of the beginning of the universe with the Big-Bang a relict model of 20th-century cosmology and does not fit today's scientific thought.

Bio-Universe Model model and Alan Rayner's principle of "natural inclusion"

In DVR universal space (which has origin in a dynamic vacuum) is the fundamental energy of the universe. This energy cannot be destroyed and cannot be created, it is just continuously transforming. All that exists in the universe, from micro to the macro scale is made out of the same energy of the vacuum.

Vacuum energy is syntropic, it has no entropy; it does not "get old". That's why proton, electron, and photon which are vortexes of the vacuum energy have such a long lifetime. Entropy starts at the scale of the atom; entropy is valid only for matter and not for the energy of the vacuum.

DVR model, where vacuum is the source of both space and the fundamental energy of the universe is close to British biologist, Alan Rayner's understanding of the fundamental evolutionary principle of 'Natural Inclusion'. In a recent summary article "What, Most Fundamentally, Are We Made Of? The Revitalising Science, Art and Spirituality of Natural Inclusion", Rayner says:

"The most realistic and personally satisfying scientific answer that I have as yet been able to find is simply this: Most fundamentally we are made of space and circulating energy in mutually inclusive, receptive-responsive relationship. And this answer doesn't just apply to us human beings, but to all natural forms, from sub-atomic to galactic in scale. It can hence be thought of as a fundamental evolutionary principle of Nature. For the sake of giving it a recognizable name, I have called this principle 'natural inclusion' and described it as 'the process through which all natural form comes into being and diversifies as a mutual inclusion of space and energy in receptive-responsive relationship'. By 'space' I am simply referring to the 'intangible void stillness' of Nature, which exists everywhere, without limit. By 'energy' I am referring to the source of all natural movement within space and around local centres of space. Such movement cannot occur without space or without the passage of what we call 'time'."

In my view, Rayner's »natural inclusion« is the important principle that science today needs to integrate into its research methodology, if it is to be truly realistic. Without having an overwhelming sense of affinity throughout the universe, science cannot progress. With analytical research methodology alone, we attempt to separate off and dissect – as an isolated objective 'part'- only a 'somewhere' within this continuity while forgetting that this 'somewhere' naturally belongs as a dynamic locality within the 'everywhere' of limitless space.

Erving Schrodinger also regarded space as the fundamental essence of the universe: "What we observe as material bodies and forces are nothing but shapes and variations in the structure of space". Rayner might seek to qualify this as 'dynamically configured structure' of space, emphasizing the point that natural structure cannot validly be treated as rigidly three-dimensional or instantaneously existent, but is necessarily four-dimensional as what he calls 'place-time'.

Schrodinger was hence probably fully aware that "natural inclusion" can be fully experienced and realized: "The world is given to me only once, not one existing and one perceived. **Subject and object are only one**. The barrier between them cannot be said to have broken down as a result of recent experience in the physical sciences, for this barrier does not exist. ... ".

Natural inclusion and today's scientific analytic research methodology are, nonetheless, fully compatible so long as the latter takes into account and does not ignore or contest the reality of the former. This is because the former is comprehensively synthetic while the latter helps to focus attention on local detail. Without applying the first one, the second one can distort our understanding of reality and be misused for technological and governmental purposes that are harmful to life and humans. Natural inclusion should be taught at universities worldwide. It will raise students ecological awareness and planetary consciousness. We are primarily humans and only secondarily belong to religions, nations, and races.

The philosophy of natural inclusion (which can be called 'natural inclusionality') has immense potential to bring more acceptance between different cultures. We cannot build vibrant, diverse communities either by removing our differences or allowing them to divide us. We can only do so through recognising both what we all have in common and how our differences enable us to live together co-creatively. Natural inclusionality is pointing towards both through its recognition of the receptive continuity of space and the dynamic distinctiveness of naturally forming boundaries.

Consciousness Research and Sadhguru's Inner Engineering

For 30 years I search on consciousness. For two years I was traveling in Asia and visiting different places, practicing different techniques. In April 2019 I was attending the course "Inner Engineering" in London lead by Sadhguru.

The experience of this course is very positive for me. Inner Engineering is the science of inner discovery of the source of life which is consciousness itself. I highly recommend this course to all scientists on search in the subject of consciousness.

Search on consciousness is primarily subjective, our scientific models of consciousness are only pointers in the right direction which is far beyond the rational realms of the scientific mind. We have to admit in the science that consciousness is ontologically far deeper than the mind. This is the first necessary step into experiential consciousness research.

The progress of human society is possible only with the awakening of consciousness worldwide. Inner Engineering is according to my experience the best tool to deepen into consciousness and should be introduced in the educational system worldwide. Inner Engineering has the potential to strengthen ties between different cultures because consciousness is the real core of every culture. Consciousness is the source of real intelligence and creativity which enriches human life and preserves nature.

Consciousness is the source of the observer. By the awakening of the observer, scientists worldwide will immensely increase their ethical potential. Nobody will be ready to participate in the research which is harmful to life. Science will start serving life and disobeying the idea of profit as the only sense of life.

On this planet, we are investing huge money into colliding protons in extremely expensive cyclotrons and building an imaginary physics of elementary particles which are man-made and absolutely have no importance for physics and for the prosperity of human society. Investing this money in the dynamic vacuum research, already we would have the technology of antigravity and the technology of energy extraction from the vacuum.

7

Time has only the mathematical Existence

The Minkowski formula X4=ict indicates that time t does not express the 4th dimension of space-time, i.e., X4 is not t. Therefore, Minkowski space-time is not a 3D+T manifold, it is 4D. The 4D manifold has 4 spatial dimensions and is timeless, in the sense that it does not contain some physical time in which material changes run. In physics, we experience the timelessness of 4D space as "Einstein's Now." From this perspective, the time that we measure with clocks is just the sequential numerical order of changes, i.e. motions that run in the timeless space of Now. On the other hand, past and future are nothing more than psychological or conceptual realities that derive from the neuronal activity of the brain. Therefore, "time flow" and "arrow of time" have only a mathematical existence.

Bijective analysis confirms the Minkowski manifold confirms that space-time is timeless. In this way, "Einstein's Now" fully enters the realm of physics, in the sense that, in physics, we experience the timelessness of space as Now. In the Minkowski manifold the formula for the 4th dimension of space-time is as follows:

$$X4 = ict \ (1),$$

the fourth dimension is the product of the imaginary number i, light speed c, and time t. This formula has an internal structure that is similar to the formula for distance d, in which the imaginary number i is missing:

$$d = vt \quad (2),$$

where d is distance, v is speed, and t is time. The formula (1) confirms that the 4th dimension of space-time is not temporal, but is also spatial, as are the other three dimensions: X_1, X_2, X_3. Thus, the space-time manifold of Minkowski is not 3D + T. Therefore, the generalization that time is the 4th dimension of space is not appropriate. Rather, the space-time manifold of Minkowski is 4D. Minkowski's vision was to fully integrate space and time as an inseparable entity. On September 21, 1908, Hermann Minkowski began his talk at the 80th Assembly of German Natural Scientists and Physicians with the now famous introduction: "The views of space and time which I wish to lay before you have sprung from the soil of experimental physics, and therein lies their strength. They are radical. Henceforth space by itself, and time by itself, are doomed to fade away into mere shadows, and only a kind of union of the two will preserve an independent reality" (Minkowski, 1952). Minkowski was right. The mathematical formalism X4=ict confirms that, in his model, time is fully integrated within a timeless 4D space.

The research of Catalin V. Buhusi and Warren H. Meck confirms that the human experience of duration (time interval) is based on the neuronal activity of the brain (What makes us tick? Functional and neural mechanisms of interval timing, Nature Reviews Neuroscience, volume 6, pages755–765, 2005).

Therefore, the linear time of "past-present-future" exists only as a function of the neuronal activity of the brain. We experience the run of changes in timeless space within the frame of the conceptual linear time of "past-present-future." Einstein himself was aware that the linear time of "past-present-future" was only a psychological reality. He used to say: "People like us, who believe in physics, know that the distinction between past, present, and future is only a stubbornly persistent illusion" (Einstein 1915). The Minkowski manifold $X4=ict$, in which time t is just an internal element of the formula for the 4th imaginary dimension, is a mathematical model that is completely consistent with Einstein's view of time.

The solution with regard to bringing the NOW into physics is through an understanding that material changes actually run in the physical reality of 4D timeless space, which we experience as "Now," and not in the psychological or conceptual reality of "past-present-future." In order to illustrate this, let us take an everyday example: walking from point A to point B we experience that our walking occurs in space, and always takes place NOW. We do not perceive with our senses that we walk in time. We have a sensation of moving in time because we experience our motion in the frame of the linear conceptual time of "past-present-future," which as previously stated, has only a psychological existence. Another example of material changes actually running in the physical reality of 4D timeless space can be found by examining the motion of a photon in space. The photon passes Planck distance in Planck time, which is just the numerical order of the photon's motion. The photon is not moving in some physical time, as elementary perception and experimental data both confirm that the photon moves only in space. When we calculate the sum of the Planck times of the photon's motion from point A to point B we get time t as the duration:

$$t = t_{P1} + t_{P2} + ... + t_{PN} = \sum_{i=1}^{N} t_{Pi} \quad (3).$$

What this demonstrates is that changes run in timeless space and that the existence of duration requires a measurement from the side of the observer. Equation (3) expresses the duration of the photon's motion from point A to point B, which is generated (and thus enters existence) as a consequence of the measurement of an observer. For this reason, we can call the time t that expresses duration the emergent time. Without a measurement, there is no duration. Moreover, by taking into account that the existence of duration of physical events requires that the observer make a measurement, one can speculate that there are two ways to understand time:

– Time measured with clocks is a numerical order of change that has only a mathematical existence;

– Duration of a given material change requires that the observer makes a measurement.

These two ways to understand time indicate that in physics we have two kinds of time:

1. *Fundamental time*, which is the numerical order of change and exists independent of the observer.

2. *Emergent time*, which is the duration of material change and originates from the observer's measurement.

Einstein's famous quote confirms this view of time: "Time has no independent existence apart from the order of events by which we measure it". The view presented above solves the Now conundrum in some detail, as follows: Fundamental time is the sequential numerical order of changes that run in the Now, whereas emergent time is the duration of changes that run in the Now, and only comes into existence when measured by an observer.

In physics we do not have any tangible experimental data that time has a physical direction, nor do we have any tangible experimental data that the physical arrow of time exists. Now is not a problem once one understands that time is the sequential numerical order of changes running in the timeless space of Now.

Shapiro has observed through measurement that in a relatively strong gravitational field light has a diminished speed. In an article entitled "Fourth Test of General Relativity" Shapiro wrote: "Because, according to the general theory, the speed of a light wave depends on the strength of the gravitational potential along its path, these time delays should thereby be increased by almost $2x10^{-4}$ sec when the radar pulses pass near the sun. Such a change, equivalent to 60 km in distance, could now be measured over the required path length to within about 5 to 10% with presently obtainable equipment".

The solution to Einstein's NOW conundrum presents us with a new physical model in which light moves in space, which is always now, and time is its duration. In this new physical model, time does not exist as a physical reality (time as duration is only an emergent quantity which originates from observer's measurements) and as such, cannot shrink or dilate. The results of Shapiro confirm that in a relatively strong gravitational field the speed of light is diminished only minimally, because in a relatively strong gravitational field what actually changes are the permeability and permittivity of space, which changes only minimally influence light speed. In a relatively strong gravitational field light is moving a bit slower and so needs more duration to move from point A to point B than it does in space with weaker gravity. Understood in this way, it is more appropriate to refer to this effect as a "gravitational velocity decrease," rather than as a "gravitational time delay." Several pieces of research already confirm that gravity changes the permittivity and permeability of space and that these changes diminish light speed.

However, this effect applies not only to light speed, because the rate of all physical changes diminishes in a relatively stronger gravitational field. In GPS systems, clocks on the satellites are calculated to run 45 microseconds per day faster than clocks on the Earth's surface, because gravity is stronger on the Earth's surface. This is the so-called "General Relativity (GR) relativistic effect." However, in GPS systems, clocks on the satellites are also calculated to run 7 microseconds per day slower than on the Earth's surface, owing to "Special Relativity (SR) time dilation," which would be better named "SR velocity decrease." Owing to this combination of GR and SR effects, the rate of clocks on the satellites is faster than the rate of clocks on the Earth's surface by 38 microseconds per day.

However, the rate of clocks on the Earth's surface does not get slower because they exist in a physical reality called time that can actually dilate. In SR and GR (and in physics in general) time is only the mathematical parameter of material changes, i.e. motion, and cannot be "relative." We see in GPS systems that what is actually "relative" is the velocity of material changes i.e. motion. All clocks are running in the same timeless space of Now. In a relatively stronger gravitational field the rate of clocks is slower, the speed of light diminishes, and the velocity of physical and biological changes diminishes as well.

As regards the famous special relativistic argument of the twins, once it is realized that the twins are both getting older in space, and not in time, there is no longer a "twin paradox." A twin on the Earth's surface ages faster than his or her counterpart on a fast spaceship owing to the SR relativistic effect. A Twin on the Moon ages faster than his or her counterpart on the Earth owing to the GR relativistic effect. The rate of the aging process for each twin is therefore different, but is real for both twins, because both twins actually only ever age in the timeless space of Now.

In a universe where the fundamental arena is timeless space, traveling into either the past or the future is categorically excluded. The idea of time travel, as something that might actually be possible, was born with the prediction of "closed timelines," or the more frequently used term, "closed timelike curve (CTC)," which were predicted by Willem Jacob van Stockum and Kurt Gödel. However, Gödel acknowledged that CTC allows for contradictive time travel, leading him to conclude that time cannot have a physical existence. He expressed this view in a famous statement: "In any universe described by the theory of relativity, time cannot exist." Stockum and Gödel simply used the wrong term to describe the "closed timelines" that they discovered, and this in turn then led them to the erroneous conclusion that one could travel in time. I show in this book that the fourth coordinate of space is not temporal, but is also spatial. Hypothetical movement in "closed timelines" means that we move only in space, and so we always end up at the same point from which we started, which is in space now. Motion in time is categorically excluded because time is merely the duration of motion in space.

Moreover, the idea of time as the sequential numerical order of changes running in the timeless space of NOW allows us to justify, in a clear way, in what sense in general relativity the idea of an idealized time t that flows on its own in the universe, without reference to anything that happens, must be abandoned and be replaced with different possible internal times associated with specific physical clocks. Taking this approach allows one to deal with the relative motion of the variables, with respect to each other, in a democratic fashion. In general relativity, there is not a preferred and observable quantity that plays the role of an independent parameter of the evolution of a system, because clocks provide only a mathematical measure of the numerical order of physical events in the timeless space of .NOW. With clocks one measures frequency, speed, and numerical order of events in the timeless space of NOW. Since clocks can be defined as those instruments which measure the speed of material changes and movements, the internal clocks/times of general relativity are only measuring systems, and what they measure is the numerical order of material changes in the timeless space of NOW. The definition of time as a mathematical coordinate that indicates the sequential numerical order of material motion running in the timeless space of Now thus provides a clear and suggestive re-reading of the nature, significance, and meaning of the internal clocks/times of general relativity.

In bijective physics, we have shown coordinate time (temporal dimension) in Minkowski manifold is also spatial. Clocks run in a timeless 4D space. The relativistic rate of clocks depends on the strength of the gravitational field in a given region of 4D space and is valid for all observers. GPS systems prove this point beyond any reasonable doubt.

Bijective physics view of time as the sequential numerical order of material changes in the timeless space of Now allows a resolution of this problem in a clear and unifying way, as follows: both in relativity and in quantum mechanics time is a mathematical parameter measuring the numerical order of material changes in the timeless space of NOW.

With respect to the arguments made in this chapter, one can conclude that the fundamental arena of natural processes is a timeless background, and that material changes run in the timeless space of NOW. This means that, when the change X_2 comes into existence, the change X_1 no longer exists; when the change X_3 comes into existence, the change X_2 no longer exists. Changes are irreversible, in the sense that a change that no longer exists is lost forever. Any increase in the entropy of a given system only happens in timeless space. In other words, "time flow" and "arrow of time" have only a mathematical existence. Time (as the sequential numerical order of changes) flows within the timeless space of Now, and has an arrow which points in the direction of numerical order increase: $X_1, X_2,...,X_n$.

During the period of time in which Einstein came up with his Relativity theory, what we now know to be the linear psychological time of "past-present-future" was still considered to be actual physical reality, and so the ability of this idea to influence physics remained strong. As a result, the idea of linear time as being the fourth dimension of space was incorporated into physics. In this way, time was sort of half-way integrated into space, somewhat like the way a round peg can be sort of half-way integrated into a square hole if one pounds on it hard enough.

In Relativity it is accepted that motion from point A to point B in space takes place in 3 spatial dimensions and one temporal dimension. In this book however, we have shown that, in relativity regime, motion happens in 4 spatial dimensions, and we have also shown that that motion happens exclusively in timeless space. In GPS systems clocks run simultaneously in timeless space. Their "relativistic rate" depends on SR and GR effects and is valid for all observers. As time is not an actual physical dimension, no clock can "tick" in some hypothetical physical past or future, since past and future, as has also been described in this article, are purely conceptual realities, and as such have no demonstrable physical correlate. All clocks in this universe run in the same timeless space, which means they all only ever run Now. GPS systems have proven that the relative rate of clocks is valid for all observers. This means that an observer on a train-station platform where there is a stationary clock, and an observer in a passing train where there is moving clock, will each observe the individual clocks to run at the same rate, i.e., they will be able to agree that the clock on the train, which is moving, is running at a somewhat slower rate than the clock on the platform. In this famous example of Special Relativity, it needs to be understood that the stationary observer and the stationary clock, as well as the moving observer and the moving clock, all exist, and so all run, in timeless Minkowski 4D space.

In 20th-century physics, time is understood as a fundamental physical reality in which the universe exists. However, up to now, we still have not obtained a single piece of data that provides any evidence whatsoever confirming time to have an actual physical existence. Understanding time to have only a mathematical existence provides a plausible solution to Einstein's NOW conundrum: time is the mathematical parameter of changes i.e. motions, which run in the timeless space of NOW.

Elementary perception (sight) and experimental data confirm we measure with clocks sequential numerical order of events which means that flow of time has only the mathematical existence. The idea of some physical flow of time has no support in elementary perception and experimental data and should be abandoned. On the other hand in today physics, we have several other models of time: coordinate time, proper time, internal time, external time, thermal time, cosmic time. None of these times we are able to observe with our senses. In this article it is shown these models of time have no real physical existence, they are pure theoretical mistakes.

Since physics exists we measure with clocks duration of material changes, i.e. motion. If we say that "time is duration" no one who knows physics can object to this statement. The common interpretation that duration exists in some physical time has no experimental evidence and we will abandon it. We can only claim that duration exists in space. You do a simple experiment: you move a pen on your table from the left to the right side of the table. You can experience only motion in space. Without measurement, the motion itself has no duration on its own. We can conclude on this simple experiment that a given motion in order to have duration needs to be measured. Time as duration exists when is measured. If there is no measurement from the side of the observer there is no duration. Time as duration is the result of the interaction of the observer with the given physical event. The following question arises: "Does time run without measurement"? We can observe in the universe that every physical event has its own sequential order. Let's take the example of photon motion from the left side of the table to the right side of the table. Photon is moving from point A to point B in space so that it passes from one Planck distance to the next Planck distance. When a photon is on the Planck distance l_{P1} it is not anymore on the distance l_{P2} and so on. Each Planck distance l_{PN} which photon is passing corresponds exactly one Planck time t_{PN}. In this sense Planck times $t_{P1}, t_{P2}, ... t_{PN}$ represent sequential numerical order of photon motion from A to B. In this view time is the numerical order of photon motion. We name this time "fundamental time" because it exists without the measurement of the observer. When fundamental time is measured by the observer "emergent time" which is duration enters existence. Emergent time in the case of photon motion from A to B is the sum of Planck times.

$$t = t_{P1} + t_{P2} + \dots + t_{PN} = \sum_{i=1}^{N} t_i$$

Planck time t_P is the unit of fundamental time which has only the mathematical existence. Our experience and experimental data confirm: time is real, time exists; time is the sequential numerical order of physical events; time has only the mathematical existence. The flow of fundamental time does not run in some physical time as the 4th coordinate of space. Fundamental mathematical time runs in space; in space is always NOW. With clocks, we measure the duration of events in space. The relative velocity of events (rate of clocks included) depends on the variable energy density of space and is valid for all observers. GPS system proved this with no doubt. Shapiro experiment proves that light has a bit lower velocity in stronger gravity (where energy density of space is lower) which means that the same photon clock (or any other clock) will have a slower rate in stronger gravity and faster rate in intergalactic space where gravity is weak and energy density of the space is at the maximum.

The "paradigm shift" in the understanding of time is the following:

- The flow of time has only mathematical existence.

- Material changes, i.e. motion run only in space and time is their sequential numerical order.

- "Relative" is the velocity of a material change, i.e. motion; time is not relative, it cannot dilate or shrink. Time cannot go faster or slower, only the velocity of physical events can be faster or slower.

- The relative velocity of material change (rate of clocks included), i.e. motion is valid for all observers and depends only on the variable energy density of space and not on the position of the observer.
When measured, every elapsed time is the multiplication of Planck time. Time is not continuous, time is discrete. We can denote this with the following formula:

$$t = t_p \cdot n \, ,$$

where n is finite natural number.

As energy also time exists in discrete quantities. The difference is energy has physical existence; time has merely mathematical existence. According to the bijective physics, the universe is non-created and eternal which means numerical order of events is infinite. This is the so-called "fundamental" time". When we measure the numerical order of an event we get duration which is "emergent" time. Emergent time is finite.

In his book "The End of Time: The Next Revolution in Physics«, Barbour is denying the existence of physical time which is right. On the other hand, he claims: "I will not claim that time can definitely be banished from physics; the universe may be infinite, and black holes present some problems for the timeless picture. Nevertheless, I think it is entirely possible – indeed likely – that time as such plays no role in the universe". I do not agree with Barbour's view, on the contrary: "Fundamental time *as such* is the numerical order of events and is at the very core of the universe despite it has no physical existence". Time is the fundamental element of the universe and fundamental element of physics. The solution is not in banishing time as Barbour suggested, the solution is to give it the right meaning.

Rovelli and Connes also are misunderstanding the real nature of time which is time has only the mathematical existence: "We consider the cluster of problems raised by the relation between the notion of time, gravitational theory, quantum theory and thermodynamics; in particular, we address the problem of relating the "timelessness" of the hypothetical fundamental general covariant quantum field theory with the "evidence" of the flow of time. By using the algebraic formulation of quantum theory, we propose a unifying perspective on these problems, based on the hypothesis that in a generally covariant quantum theory the physical time-flow is not a universal property of the mechanical theory, but rather it is determined by the thermodynamical state of the system ("thermal time hypothesis" (A. Connes, C. Rovelli, »Von Neumann Algebra Automorphisms and Time-Thermodynamics Relation in General Covariant Quantum Theories«, Class. Quantum Grav. 11:2899–2918, 1994). My comment here is that physical time-flow cannot be determined by the thermodynamical state of the system because physical time-flow does not exist. Time is the numerical order of the changes of a given thermodynamic system.

Italian physicist Claudio Borghi is introducing "internal time" and "external time": "A careful analysis shows that in physics the concept of time is used in two different ways: as an external attribute of motion or as an implicit variable that measures the internal evolution of a system. The first one is explicitly used in mechanics, the second, implicitly, in thermodynamics. Since in thermodynamics the variable t in practice does not appear in the definition of the physical quantities, we naively think that the concept of time introduced in mechanics can be used everywhere. Although it may sound a simplification, it is immediately clear that, as the mechanical evolution is related to the change in the position of a body with respect to others, the thermodynamic evolution of a system is linked to processes that involve it internally and might not have relationships with the environment, thus with the external space of relations" (Annales de la Fondation Louis de Broglie, Volume 37, 2012).

Borghi introduction of internal time and external time seems not necessary as time is the mathematical parameter of motion of space and the mathematical parameter of a thermodynamic evolution of a given system.

Einstein has introduced in Special Relativity "coordinate time" and "proper time" of a given inertial system. The 4th coordinate of Minkowski manifold is not "time coordinate", it is spatial too because $X_4 = ict$ means the product of velocity c and duration t which is spatial distance. Minkowski manifold is not $3D + T$, it is $4D$. In this 4D manifold time has only the mathematical existence as we have seen in the introduction. Einstein "coordinate time" model has no bijective correspondence in the physical world.

The common view in cosmology is that some "cosmic time" exists as the time coordinate since the big bang. No one ever has seen this cosmic time and we have no experimental data confirming its existence. In this article is proposed that the term "cosmic time" is abolished because we have not a single evidence of its existence. The universe is running in space (in which is only and always NOW) where time is merely the sequential numerical order of universal changes.

Physicist Lee Smolin is arguing in his book "Time Reborn" that physical time is a physical reality in which universe runs. He does not give any experimental prove about the physical existence of time. Nobody ever gave it any proof, so would be wise we take into account in physics that physical time does not exist.

Isaac Newton said:" Hypotheses should be subservient only in explaining the properties of things but not assumed in determining them, unless so far as they may furnish experiments". We have many hypotheses of time and none of them has experimental verification. By giving them credibility we are not ensuring to progress, on the contrary confusion about what is time is increasing. On the basis of elementary perception (sight) and experimental data we can conclude that flow time is real, time exists, but it has only the mathematical existence. It is time we acknowledge in physics that also non-material things exist. Time is one of them.

Considering time flow has only the mathematical existence hypothetical time travels are categorically excluded. We can travel only in space in which is always NOW. Twin brother on the Moon aging will be faster than the aging of his brother on the Earth, but both are getting older only in space. Hypothetical travel through wormholes in some "other time" is out of the question. "Time Machine" hypothesized by some researchers is out of the question. Linear time "past-present-future" exists only in the human mind as the psychological time through which we cannot travel with the spaceship. "Symmetry in time" is another model which does not correspond to the physical reality. A given physical phenomenon can only be symmetric in space because physical time is nonexistent.

Time arrow, outer and inner research

We have in physics today several models of "time arrow". Eddington said back in 1928: "The thermodynamic arrow corresponds to the direction of increasing disorder and hence of entropy. Thus it is also called the entropic arrow of time. It follows from the second law of thermodynamics. Time proceeds in the direction of the increase of entropy".

The Cosmological arrow of time is the direction in which the universe is expanding at present. The quantum mechanical arrow of time is the direction in which wave function collapse or state vector reduction occurs leading to definite state from among many possibilities. The psychological arrow of time is the direction from past to future that we assign to perceptual time because we remember the past and not the future.

None of the models of time arrow one can perceive by senses in physical reality or measure them by instruments. The only existent "arrow of time" is a psychological one. The observer which is locked in psychological time will experience the change X_1 is entering existence after the change X_1, the change X_3 is entering existence after the change X_2 in some physical time which he cannot perceive by senses (because is nonexistent). The observer which is aware of inner psychological time will experience the flow of material changes in space in which time runs only as the numerical order of these changes; when change X_2 is entering existence, change X_1 does not exist anymore. When change X_3 is entering existence, change X_2 does not exist anymore. The entire universe is dying and being born in the eternal moment of NOW. This perspective of time is the necessary tool for the advancement of physics.

Psychological arrow of time has a physical origin in the neuronal activity of the brain. The observer can reach beyond the brain by "watching", "witnessing" they way your mind functions, which is giving you the experience of the conscious observer which is beyond the matter and beyond the mind. The conscious observer can watch the way his mind functions.

Regarding the physical existence of time Albert Einstein, was "out-of-the-box" thinking. He said: "People like us, who believe in physics, know that the distinction between past, present, and future is only a stubbornly persistent illusion".

Albert Einstein was the embodiment of mysticism and physics in one person. He said: "I maintain that the cosmic religious feeling is the strongest and noblest motive for scientific research".

The same is valid for Max Planck: "All matter originates and exists only by virtue of a force... We must assume behind this force the existence of a conscious and intelligent Mind. This Mind is the matrix of all matter".

Today's physics is locked in reductionist rationalism which is suffocating the progress of physics. Mathematic has overruled physics and phenomena are discovered for which we do not have direct experimental evidence.

Einstein said: "I do not believe in mathematics". What he meant is that mathematics is a tool which we use in building models of reality, but we can not build these models exclusively on mathematics, rather we build them on elementary observation. By applying bijective research methodology one is sure that his model represents an adequate picture of physical reality. The bijective research methodology is based on the observation of a given element (phenomena) in physical reality. The observed element has exactly one correspondent element in the model of reality.

Today "peer review" is considering that a given article has scientific validity if there is enough mathematical modeling. Nobody is posing the question if the phenomenon which is mathematically described, has bijective correspondence with the physical world and is falsifiable. "Bijectivity research methodology" and Carl Popper "falsifiability" are two tests which are drawing a clear line between science and pseudoscience.

A given scientific model is falsifiable if it is possible that it could be wrong. In science, we do not have absolute truths. When it is possible that a given model could be wrong, only then we can improve it. In today science has happened that Karl Poper vision of science is forgotten. Higgs mechanism, for example, is not falsifiable.

Karl Popper said: "A theory should be considered scientific if, and only if, it is falsifiable". A scientific statement is one that could possibly be proven wrong. Such a statement is said to be falsifiable. The falsifiable statement always remains tentative and open to the possibility that it is wrong.

In the Physics Hypertext Book we can read following regarding the Higgs mechanism: "All of space is assumed to be filled with a Higgs field — a background sea of virtual Higgs bosons that pop in and out of existence. The quarks, leptons, and W and Z bosons moving around through space interact with this field, which is why these particles have mass. The photons and gluons do not interact with the Higgs field, which is why these particles do not have mass. Even the Higgs boson itself interacts with the Higgs field. It gives itself mass!".

All of space is assumed to be filled with a Higgs field — a background sea of virtual Higgs bosons that pop in and out of existence. This statement is not falsifiable and cannot be considered scientific.

The quarks, leptons, and W and Z bosons moving around through space interact with this field, which is why these particles have mass. This statement is not falsifiable and cannot be considered scientific.

The photons and gluons do not interact with the Higgs field, which is why these particles do not have mass. This statement is not falsifiable and cannot be considered scientific.

Even the Higgs boson itself interacts with the Higgs field. It gives itself mass! This statement is not falsifiable and cannot be considered scientific.

The mainstream science has in a kind of religious belief that »discovery« of Higgs boson is a scientific confirmation for the existence of the hypothetical Higgs field. If you pose any sensible question regarding the existence of Higgs field, you are out of the game, you are dissident, you are a disbeliever. It is incredible how the scientific mind which has no awakened observer as a supervisor, can go astray and build pseudoscience. Higgs mechanism is pseudoscience.

There is nothing wrong with scientific reasoning and logic. I use them daily in my research. Wrong is that scientific logic is now destroying its own fundaments. Higgs mechanism has brought in physics irrationalism which is prooved with the fact that CERN has no answers on the most sensible questions regarding the Higgs mechanism.

I sent to CERN physicists a few kind letters where I pointed out that introducing a new field in physics in order to describe the mass of the elementary particles is a big theoretical mistake because the massive particle has its mass accordingly to the amount of its energy. For Einstein »mass« and »energy« are the same »stuff« in different forms. With Higgs mechsanism, »mass« has become that nobody understands anymore. Just see a few videos about the Higgs mechanism and you will see that nobody understand the subject. They even do not distinguish between »mass« as the amount of energy and »inertial mass« as the physical property of the proton which has origin in the proton interaction with the vacuum. Higgs mechanism is prooving that today scientific mind is totally lost in his own labyrinths.

CERN physicists never answered to me because they have no answer. Higgs mechanism has no power, to give answers on the most sensible questions regarding the origin of mass of elementary particles. This is quite a sad story.

There will be no progress of theotritcal physics until in the process of education we do not develop a conscious observer which is the guardian of the scientific mind. The conscious observer is fully taking into account falsifiability and bijectivity and is building models of reality which are its 100% adequate picture.

The Higgs mechanism is the school example of flawed methodology where something is theoretically predicted, mathematically described and finally "discovered" by an indirect experiment is destroying physics beauty and exactness. It will be recognized soon that observation and experimental data are the pillars of physics. Mathematic is just a descriptive tool.

The most known representative of scientific logic which is lost in his labyrinth is English biologist Richard Dawkins which is continuously fighting against god and glorifying scientific logic. Scientific logic is a good tool to search a part of reality which we can reach by elementary perception.

With senses, we are reaching about 5% of reality. Pretending that 95% reality which our senses cannot reach does not exist is a big ignorance. Richard Dawkins is the representative of this ignorance which has put humanity in today unbearable condition where we are destroying this planet in the name of profit.

Outer research of the outer material world and the inner research of inner psychological world and spiritual world are complementary. In both the observer is the central point. The same observer is exploring outer world and inner world.

Once you are entering the state of conscious observer you are aware the way your mind creates models of reality. Then you have better possibilities to build a model of reality which will have bijective correspondence with the reality itself. Research on time is the classical example that without inner research and discovery of conscious observer we will never understand time. The conscious observer is beyond matter and beyond the mind. Its source is consciousness itself. It is the noblest discovery one can achieve. Albert Einstein said: "The most beautiful thing we can experience is the mysterious. It is the source of all true art and all science".

Careful examination of different models of time shows that all these models are not built on perception and experimental data, they are pure theoretical speculations. Physics today is in deep crisis because today is modern to invent new models and publish theoretical articles about them. In this way, we will not have any progress. The progress of physics is building models of the world which are based on elementary perception and proved by the experiment. The spirits of Isaac Newton, Albert Einstein, Karl Popper, and Max Planck should be respected in order to progress research on time and physics in general.

Bijective physics flows following research methodology:

1. Observation of the studied phenomenon.
2. Development of the model where the phenomenon is

described in a bijective way.

3. Experimental verification of the model where results are directly interpreted.

Today in physics results are often interpreted in the way we like to understand them. For example, diminishing of the orbital velocity of the binary neutron stars PSR B1913+16 has been interpreted as the proof for the existence of gravitational waves. Diminishing of the orbital velocity is due to the diminishing of the mass. The idea that mass in transforming in gravitational waves is an unproven hypothesis. According to my calculations on the surface of binary stars PSR B1913+16 the density of the vacuum is so low that matter disintegrates back into the primordial energy of the vacuum. Sure this process could create gravitational waves, but we do not have experimental data proving hypothesis yet.

Also, LIGO results are interpreted. What is measured in LIGO is that the velocity of light is changing when passing in the beams. Now, the interpretation of this data is, that gravitational waves are causing this variability of the light speed, namely, gravitational waves should stretch and dilate space, so light needs more or less time when passing the beams of the interferometer. What we measure directly is light speed is changing and the only possible direct explanation is that permittivity and permeability of the vacuum are changing. If gravitational waves are the cause of permeability and permittivity change remains an open question.

Dynamic Vacuum Relativity **DVR** model works well without length contraction and time dilation. Actually, length contraction was introduced by Hendrik Lorentz in order to save "ether theory". Einstein has used the length contraction in his **SR** and **GR**. In **DVR** I drop these two models as they do not have bijective correspondence in the physical world. Nobody ever has measured objects would shorten along with their motion. Nobody ever has measured that time as the 4th dimension of space is dilated. It is time we replace in physics the models of length contraction and time dilation with the model of the variable density of the vacuum which is the actual physical cause of variable velocity of material changes, with the rate of clocks included.

8

New Paradigm includes Research on the Origin of the Observer

Everything in the universe is changing and improving. Today, science is stubbornly against this process despite data which are clearly proving that old theories are dead. Astronomical observation confirms that the observed universe is 10000 bigger than the universe according to the big-bang model. It is proved without any doubt that time has only the mathematical existence and CMBR radiation cannot have origin in some physical past (which is nonexistent). NASA has confirmed that universal space has a Euclidean shape and is infinite. Still mainstream is sticking to the big-bang.

The idea of evolution as the accidental process where the fight for survival and random mutation are pushing the evolution of life is (beside big-bang model) the most primitive idea of today science for which we do not have a single experimental data and still, Darwinism is meant to be scientific knowledge. Evolution of life is the consistent part of universal dynamics and runs in the entire universe. And we are still sticking to the geocentric view in biology which sees life is developing on the Earth surface. This is relict thinking from the times we have been thinking Earth is flat.

Particle physics is holding on the idea of particles are existing in an "empty space" deprived of physical properties. This conviction is preventing physics to unify quantum physics, quantum electrodynamics, Relativity Theory, and Higgs mechanism.

In today's science, we still do not understand that the observer which is a crucial element of science is not stored in the brain. The observer is subjective and can be searched only in the subjective way by watching, witnessing the way the scientific mind works. We are holding on the idea of "quantum consciousness" naively thinking that consciousness is made out of "quanta".

The only path out of this labyrinth of false convictions is that we actively start watching the way the scientific mind works in order to confirm his old beliefs. It is time tom wake up and leave behind the past ideas which have been built on reductionist convictions rather on experimental data. We need a new scientific paradigm where the observer will be searched as the core of physics. The search of the observer is the ground stone of future science which will protect nature and life and fully support the development of human society. Dear reader, I hope this book has refreshed your mind and you are seeing the universe, life, and yourself in a clear mental vision. We are not lost in some desperate universe which will expand forever and end in "thermal death". We are living in the universe which is an eternal non-created intelligent system where the human being is only a small flower. The aim of this book is that this flower blossoms fully and becomes the source of eternal fragrance for you and your beloved.

9
Bijective Economy and Bijective Banking

The entire universe obeys the first law of thermodynamics: energy cannot be created, it cannot be destroyed, it can only be transformed into another type of energy. In the universe, and in nature there is no principle of the "profit", apple is growing to become food for animals and humans, apple is not growing in order to be sold and create profit. In the universe and in nature, energy is freely flowing and is in constant transformation. Every tiny organism in this chain of energy transformation plays an important role. Without bees life on the planet cannot survive. It cannot survive without worms, without microorganism in the soil which we are unknowingly killing with pesticides in order to make a bigger profit.

The idea of the profit is a man-made idea and does not have a place in the universe and in nature. In-universe and nature energy is freely flowing without the intent of making a profit. In Bijective Economy, you can make money only with working. Work means adding energy to human society in different forms of professions. In physics, energy is measured by Joules, in the economy, energy is measured by money. In Bijective Economy, you cannot make more money by turning the money around, means with "investing". In Bijective Economy in order to make money, you need to work.

Making money with the investment is against the first law of thermodynamics. World stock market is making trillions of fictive money which value is maintained by the daily work of the people. That's why we always have inflation. Inflation is the result that someone is taking money from the system by making a profit on the stock market. When this robbing of the money reaches a huge dimension super-inflation happens and eventual collapse of the financial system as it was the case back in 1930.

The bijective economy requires abolishment of the stock market worldwide which is an organized system for taking the money from small investors. The bijective economy requires that money is backed with the gold in the national banks. Banks should be strictly owned by the state which is the guarantee that the banking system works for the citizens. Banks should not function to make a profit, they are generators of the free flow of the money which is the blood of human organism society.

Such an economy and banking system are the base for the solid and prosperous development of society. Sure bankers will not agree with this idea. However, this is the necessary change we need in order to come out of the world crisis of continuous wars, pollution, and increase of the entire population sickness.

Today, the disease is the source of the profit and the entire society works in the direction of increasing the number of ill people. We are chemically polluting soil, air, underground water, rivers, we are producing poisoned food and spending a huge amount of money for public health. Nobody is ready to switch the parading and organize society in the direction of increasing health. More health means less illness, health means income for the individual and for the state. Illness means spending money for the individual and for the state. The only ones who are making money are pesticide and pharmaceutical companies. This cannot go forever because the state has no unlimited sources and we are facing the collapse of the financial system worldwide.

Production of healthy food without pesticides should be the primary task does every state. Austria has entirely prohibited the use of glyphosate. I hope that other states will follow. This is the most intelligent action in order to preserve this beautiful planet habitable for the next generation. The time is running out. We need to wake up, otherwise, we will be eliminated by nature itself. Life on the planet will recover because life has the origin in consciousness which is far beyond chemical pollution and will generate life again.

Humans, we are not the crown of the evolution, we are "potential" which in order to come into realization needs a wider vision of a life beyond the profit in dynamic harmony with the universe and nature.

In most parliaments all over the world are making decisions which are suggested by bankers, war industry, pharmaceutics industry, and pesticides industry. Saddam Hussein and Mohammed Gaddafi were killed in the name of profit.

Because of this, the Mediterranean has become geopolitical unstable. How far this irrational policy which has the support of European parliaments will go?!

The owners of multinational companies are deciding our future. I'm sure after reading my book the light of intelligence will enlighten their minds. The purpose of human life is to make our lives better and more beautiful. With all the money which they have, they could transform this planet onto paradise. But they only think about how to make more profit. The will for the profit is the main sickness of today humanity. In nature and in the universe there is no trace of profit. All is running smoothly in the direction of an enlightened human being. Our civilization is stuck in the dogmatic thinking of the "profitable economy" which is the main cause of the poverty and stagnation of our society. The main economic resources on this planet should be invested in the development of the economy which respects cosmic laws (energy cannot be created and can't be destroyed). We need to start teaching in schools worldwide the new **economy of dynamic equilibrium** where you can make money only with work. We need our youth generation to be free of the misconceptions of the past millennium economy. Only people with a clear mind free of all superstitions can add to the progress of our civilization.

Dear reader, your comment on my book on Amazon is most welcome. It will give my book the power to bring more light and creative inspiration into today physics and into today society. For deepening into bijective physics you are invited to visit our Bijective Physics Institute (BPI) home page.

We are open to people who are willing to participate as researchers. Ph.D. is not the only requirement. You primarily must have a fresh independent mind. Our main research projects are antigravity and energy extraction from the dynamic vacuum. I'm convinced that Nicola Tesla ideas can be brought to the technological application. I also invite entrepreneurs to financially support our institute main task: "Science for the prosperity of the human race".